2nd Edition

Essential Anatomy Dissector:
Following Grant's Method

2nd Edition

Essential Anatomy Dissector:
Following Grant's Method

John T. Hansen, Ph.D.

Professor and Associate Dean
Department of Neurobiology and Anatomy
University of Rochester
School of Medicine and Dentistry
Rochester, New York

 LIPPINCOTT WILLIAMS & WILKINS
A **Wolters Kluwer** Company
Philadelphia • Baltimore • New York • London
Buenos Aires • Hong Kong • Sydney • Tokyo

Editor: Betty Sun
Managing Editor: Eric Branger
Marketing Manager: Aimee Sirmon
Project Manager: Susan Rockwell
Art Director: Armen Kojoyian
Compositor: Graphic World
Printer: QW

The publisher is not responsible (as a matter of product liability, negligence or otherwise) for any injury resulting from any material contained herein. This publication contains information relating to general principles of medical care which should not be construed as specific instructions for individual patients. Manufacturers' product information and package inserts should be reviewed for current information, including contraindications, dosages and precautions.

Printed in the United States of America

Library of Congress Cataloging-in-Publication Data

Hansen, John T.
  Essential anatomy dissector : following Grant's method / John T. Hansen.—2nd ed.
    p. cm.
  Includes index.
  ISBN 0-7817-3283-2
  1. Human dissection—Laboratory manuals. I. Title.
  QM34 .H26 2001
  611—dc21

                                                                    2001050349

The publishers have made every effort to trace the copyright holders for borrowed material. If they have inadvertently overlooked any, they will be pleased to make the necessary arrangements at the first opportunity.

To purchase additional copies of this book, call our customer service department at **(800) 638-3030** or fax orders to **(301) 824-7390.** For other book services, including chapter reprints and large quantity sales, ask for the Special Sales department.

For all other calls originating outside of the United States, please call **(301)714-2324.**

Visit Lippincott Williams & Wilkins on the Internet: *http://www.lww.com.* Lippincott Williams & Wilkins customer service representatives are available from 8:30 am to 6:00 pm, EST, Monday through Friday, for telephone access.

00 01 02 03 04
1 2 3 4 5 6 7 8 9 10

PREFACE

Grant's Dissector initially appeared over half a century ago in 1940 and has been one of the premier dissectors available for human anatomy courses in the health professions schools. Although the approach used in Grant's Dissector is well known and time-honored, medical and dental school curricula have changed significantly since 1940, not only in their emphasis but also in the time allotted for anatomical dissection. In response to the need for a more concise dissector that maximizes the students' actual laboratory dissection time, the first edition of *Essential Anatomy Dissector* was conceived to facilitate the actual physical work of dissection, yet highlight the essential structures that all generalist physicians should know.

Health professionals, in the limited time available for their undergraduate professional training, cannot learn all the anatomy there is to learn. Therefore, this dissector focuses on the essential anatomy and key concepts that most practitioners would agree is important to learn. In addition to identifying key structures, emphasis in the laboratory also is placed upon appreciating surface anatomy, relationships of important structures, cross sectional anatomy, and clinical relevance of the anatomical structures. To this end, the use of an appropriate atlas of human anatomy in conjunction with this dissector is critical. **Likewise, it is important that the faculty and students understand that this dissector is not meant to replace or substitute for an atlas and a good clinically oriented textbook of human anatomy.** The purpose of the dissector is to assist and direct students in the completion of the dissection in a timely manner and highlight important structures that they should find. The student is then directed to the atlas and textbook to enhance their knowledge of these structures and read about their relevance to clinical medicine. Most of this reading needs to occur outside of the dissection laboratory and is an important aspect of developing life-long learning skills in our students.

The dissector is organized so students can easily follow the dissection protocols, which are highlighted in color to separate these instructions from the remaining text. The format is in an outline style so students can easily group concepts or regional anatomy into blocks of information that are systematically presented. Important structures to be identified are highlighted in **bold** and the approach is regional, following the general procedures that have made Grant's method the benchmark of anatomical dissection. Figures and tables are used as regional guides to the dissection; however, it is expected that students will access an atlas during their dissection so not every dissection field is displayed. Moreover, *Essential Anatomy Dissector* is keyed to four of the more popularly used atlases of anatomy, making it widely applicable to individual course needs. It also is keyed to *Acland's Video Atlas of Human Anatomy.* Finally, Learning Objectives and Key Concepts at the beginning of major regional dissections help guide the students to their atlas, textbook, and the cadaver, emphasizing the important anatomical material that each student should learn regarding that region. Naturally, individual course directors may wish to supplement these objectives or concepts with expanded material of their own.

Because of the unique nature of individual anatomy courses, the dissector is written so students may begin their dissections in virtually any region of the body. In my experience, most courses begin either on the back or thorax, where the guiding principles of segmentation and bilateral symmetry are clearly evident. This dissector also should meet the needs of those courses that use some prosected material. Because the actual descriptive material is kept at a minimum, the students can use the dissector as a guide to identifying the relevant structures and then return to their textbooks for a more complete description of the anatomy. Every effort has been made to make *Essential Anatomy Dissector* a focused, directive, and concise manual of dissection that will optimize limited laboratory time, encourage student reading in a textbook, yet include the "essential" anatomy that all generalist physicians should encounter and learn, regardless of future specialization.

John T. Hansen

ACKNOWLEDGMENTS

I am indebted to all my former and current students and faculty colleagues who have taught me far more than I have ever taught them. Their honest and constructive feedback is always helpful and enlightening, and their inspiration keeps me going. Medicine is in good hands.

Thanks also to the reviewers of the second edition of the *Essential Anatomy Dissector.* Their constructive comments and keen eye for detail have corrected mistakes of the first edition.

At Lippincott Williams & Wilkins, the dedicated staff is professional at every level, and it has been a joy and a learning experience just to know them. Any mistakes are mine, just as any successes are ours. I sincerely appreciate and gratefully acknowledge the efforts of Betty Sun, Executive Editor; Eric Branger, Managing Editor; Aimee Sirmon, Marketing Manager; and Susan Rockwell, Production Manager. It is a distinct pleasure to be associated with Lippincott Williams & Wilkins, the premier publishing house in the field of human anatomy.

Because this manual follows Grant's Dissector, albeit in a more concise and focused format, I also am indebted to Eberhardt K. Saurland, M.D., who has enriched all of us by his masterful editing of *Grant's Dissector* since its seventh edition.

Most of all, I want to thank my wife, Paula, and my children for their unconditional love and support over the years. Because we've shared so much, this effort, like all the others, was multiauthored.

CREDITS

Illustrations listed below are reproduced from the following Williams & Wilkins publications.

Agur AMR. Grant's Atlas of Anatomy, 9th ed., 1991.
Figures 1.2, 1.5, 2.6–2.8, 2.12, 2.16, 2.17, 3.6, 3.9, 4.2, 4.4, 6.7C, 6.8B, 6.10 bottom, 7.12, 7.13A, 7.15, 7.17, 7.20, 7.23, and the illustrations that appear in Tables 7.2 (bottom illustration only) and 7.6 (left illustration only).

Basmajian JV, Slonecker CE. Grant's Method of Anatomy, 11th ed., 1989.
Figures 2.2 and 5.2.

Moore KL, Agur AMR. Essential Clinical Anatomy, 1995.
Figures 1.1, 1.3, 1.4, 1.6–1.9, 2.1, 2.3–2.5, 2.9–2.11, 2.13–2.15, 2.16, 2.18, 3.1–3.5, 3.7, 3.8, 3.11, 3.13, 3.17, 4.1, 4.3, 4.5, 4.18, 5.1, 5.2–5.10, 6.1–6.9, 6.13, 6.15, 6.17, 7.1–7.11, 7.14–7.16, 7.18, 7.20–7.22, 8.13, 8.19, 9.2, 9.12, and the illustrations that appear in Tables 7.1, 7.2 (top two illustrations only), 7.4, 7.5, 7.6 (right illustration only), and 7.8.

Figure 7.1 is redrawn from Wilson JL. Dissection Manual, 2nd ed. New York: Igaku-Shoin Medical Publishers, Inc., 1988:187.

Figures 2.13, 3.2, and 7.8 are redrawn from Glasgow EF, Dolph J, Chase RA, Gosling JA, Mathers LH, Jr. Clinical Anatomy Dissections. St. Louis: Mosby, 1996.

All tables that appear in this book are reproduced from Moore KL, Agur AMR. Essential Clinical Anatomy, 1995.

CONTENTS

INTRODUCTION

PURPOSE

Dissection is vital to a full appreciation of human anatomy. In no other format can students better learn the three-dimensional features of the human body, the inherent variability in structure, the importance of careful observation, the feel and texture of the dissection experience, the need for effective teamwork, or the exposure to the normal and abnormal that is part of every dissection exercise. To make the most of this experience, follow carefully the guidelines and expectations of your instructors, and always approach dissection with the utmost reverence and respect for the persons who have donated their bodies for medical education and research.

TECHNIQUE

Your instructors will provide guidelines regarding the proper care of the cadaver or specimens you work with, what instruments they wish for you to use, and the appropriate rules of conduct in the dissection laboratory. Additionally, they may alter the sequence of dissections listed in this manual and supplement or modify the Learning Objectives and Key Concepts to meet the needs of their individual courses.

The *scalpel* is used mostly for skin incisions, reflecting the skin, and in cutting large structures where indicated in the dissector. Blunt or fine dissection should be done with your *scissors, probe,* and *forceps.* Blunt dissection is best performed by using your fingers or the scalpel handle to gently separate structures. The "scissor technique" also may be used to spread structures apart or to dissect along vessels or nerves in parallel with their course. Your staff will demonstrate these techniques to you.

In most instances, when the dissector directs you to reflect the skin, you will reflect both the skin and the underlying subcutaneous tissue (tela subcutanea), unless your instructor prefers to leave the tela in place to identify cutaneous nerves and superficial veins. However, for the most part, the manual will point out important cutaneous nerves and vessels as you encounter them and encourage you to read about their distribution in your textbook and atlas. Time in most laboratories will not be devoted to the tedious dissection of these superficial structures.

TERMINOLOGY

From your textbook and atlas, become familiar with the universally approved position of the body known as the "anatomical position." Also, become familiar with essential terms such as **superior (cranial), inferior (caudal), anterior (ventral), posterior (dorsal), coronal, sagittal, and transverse.** Also understand what is meant by structures lying "superficial" or "deep" to one another.

ABOUT THIS MANUAL—A GUIDE FOR STUDENTS

Use the Learning Objectives and Key Concepts to guide your study of important "general" points that you should understand regarding the regional anatomy being learned. For the most part, reading your clinically-oriented textbook will be invaluable in achieving the intent of these objectives, and this reading is an important part of developing life-long learning skills. The concise nature of the *Essential Anatomy Dissector* precludes its use as a comprehensive source of anatomical knowledge.

The organization of the dissector is in outline format to facilitate an appreciation for regional organization and for ease of use. The dissection exercises follow the time-honored method of *Grant's Dissector,* but are abbreviated or modified in some regions to highlight only the "essential" material or to render the dissection less destructive. Key structures are highlighted in **bold** and may be required "identifications" by your instructor. Likewise, the actual dissection instructions ("cut, incise, reflect, etc.") are highlighted in color for easy identification.

The figures are intentionally kept to a minimum and are guides to the process of dissection. All students should become accustomed to using a good atlas to visualize the highlighted structures and study their important relationships. Consequently, this manual is keyed to four different atlases of human anatomy, with key illustrations indicated by a colored abbreviation and the **figure** (*Grant's Atlas of Anatomy* and *Clemente Anatomy*), **plate** (*Netter Atlas of Human Anatomy*), or **page** (*Rohen Color Atlas of Anatomy*) number of the atlas. The abbreviations include: **G:** Grant's Atlas, 10th edition; **C:** Clemente Anatomy, 4th edition; **R:** Rohen Color Atlas, 4th edition; and **N:** Netter Atlas, 2nd edition. Regions also are referenced to the Acland's Video Atlas as an abbreviation (**A**), tape number (1-5), and inclusive start-stop time index (for example: **A2. 1:43:10-1:55:30**).

In the text of the dissector, emphasis is placed on the action of muscles on the joints they cross. For those courses that place an emphasis on learning the origins and insertions of major muscle groups, tables are provided to summarize this information.

Finally, each regional dissection begins by directing students to examine the surface anatomy on either their cadaver, each other, from an illustration, or on themselves. This is a good practice to cultivate as surface anatomy is vitally important in medicine; again, a good clinically-oriented textbook or atlas can be an invaluable resource. Bony features also are important to learn and one should have access to skeletons and normal radiographs. Each regional section ends by encouraging students to study cross sections and CT or MRI films. This is important as radiographic anatomy will be how many of you will "visualize" your patient's anatomy during your medical practice.

THORAX

I. THORACIC WALL

Learning Objectives

- Throughout Chapter 1, identify structures in bold print unless instructed to do otherwise.
- Identify the features of the ribs and sternum and important surface landmarks on the anterior thoracic wall.
- Diagram the lymphatic drainage of the breast and understand its importance in the spread of cancer.
- Describe the function of the pectoral muscles and the role of the intercostal muscles in respiration.
- Describe how intercostal neurovascular bundles radiate around the thoracic wall.
- Know why the sternal angle of Louis is an important surface landmark.

Key Concepts

- Sternal angle of Louis as an important surface landmark
- Muscles involved in respiration
- Distribution of intercostal neurovascular bundles

A. Introduction. The thorax contains and protects the heart and lungs. The thoracic wall is comprised of the vertebrae, ribs, sternum, and muscles. Examine the thoracic vertebrae and note whether your cadaver has any abnormal curvatures of the thoracic spine such as kyphosis (hunchback) or scoliosis (lateral curvature).

Kyphosis - hunchback
scoliosis - lateral curvature

B. Bony Landmarks.   A3 35:46-46:00/ G1.13, 4.7/C153, 657-661/ R184, 186/ N 143, 171-172

1. **Sternum.** Identify the **manubrium, body,** and **xiphoid process.** The xiphoid process (Gr. xiphos, sword) is cartilaginous in youth but ossifies by midlife.

   The first seven ribs are connected anteriorly to the sternum by a bar of hyaline cartilage. The costal cartilages of ribs 8, 9, and 10 reach only as far as the cartilage superior to it. Ribs 11 and 12 are "floating" ribs.

2. **Ribs.** Identify the **head, neck, tubercle,** and **body.** The lower border of each rib shelters the intercostal neurovascular bundle (nerve, artery, and vein). Observe that the head of each rib articulates with two vertebral bodies and the intervening disc. The tubercle articulates with the transverse process of the vertebra with the same segmental number while the head articulates both with the vertebra above and at the same segmental level (e.g., the head of rib 5 articulates with vertebrae T4 and T5, and the tubercle with the transverse process of T5).

3. **Thoracic vertebra.** Identify the **body** and protective vertebral arch made up of the **pedicles, laminae,** and the **spinous process. Articular processes** project superiorly and inferiorly.

4. Note important surface landmarks, the **jugular notch** and **sternal angle (of Louis).** The sternal angle marks the junction of the manubrium with the body of the sternum and the point where the second rib articulates with the sternum. This important surface landmark is used to count ribs and intercostal spaces. Also note the outline of the **clavicle, acromion** of the scapula forming the point of the shoulder, the **xiphoid process,** and the location of the **nipple,** which is associated with the T4 dermatome in males (somewhat more variable in females).

   Make the skin incisions shown in Figure 1.1. (Disregard these directions if you have already dissected the Upper Limb, Chapter 6). With the cadaver in the supine position (face up), make a cut from A to the xiphoid process B. Then cut from A to C along the clavicle and around the upper arm. Extend your incision from B to C, encircling the nipple. Finally, cut from B to D extending your cut to the midaxillary line. Reflect the skin and tela subcutanea (subcutaneous tissue or superficial fascia) until the underlying muscles are encountered, but leave the subcutaneous tissue of the female breast in place. Deep to the tela lies the deep fascia, which envelops the underlying muscles.

C. Female Breast. If you have already dissected the breast and pectoralis muscles as part of the Upper Limb dissection, skip to the section on the intercostal muscles (Section D-3).

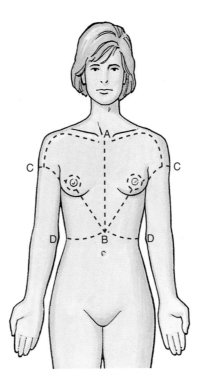

FIGURE 1.1   Incision lines.

If you have a male cadaver, join a table with a female to study the female breast. The two mammary glands lie within the tela (superficial fascia) and really are modified sweat glands. Between the gland and deep fascia investing the pectoralis major muscle lies the **retromammary space,** which allows the breast to move freely over the muscle. The gland is attached to the skin by fibrous septa, the **suspensory ligaments** (Cooper's ligaments)   **A3 1:21:31-1:22:58/ G1.3, 1.6/ C6-10/ R245/ N167-169**

1. Identify the **nipple** and **areola.** The nipple usually overlies the 4th intercostal space (may vary depending on size) and demarcates the 4th thoracic dermatome. Observe that the **suspensory ligaments** form septa that divide the breast tissue into compartments that contain fat.

2. Placing tension on the nipple, note the taut suspensory ligaments (of Cooper). With the rounded handle of your scalpel, scoop out fat from several compartments to better visualize the septa. Cut through the nipple and find several of the **lactiferous ducts** that converge on the nipple (often too small to adequately identify grossly).

3. Review on your own the important lymphatic drainage of the breast in your textbook and atlas, noting especially the drainage to the axillary lymph nodes (Fig. 1.2).   **G1.5/ C11/R245/ N169**

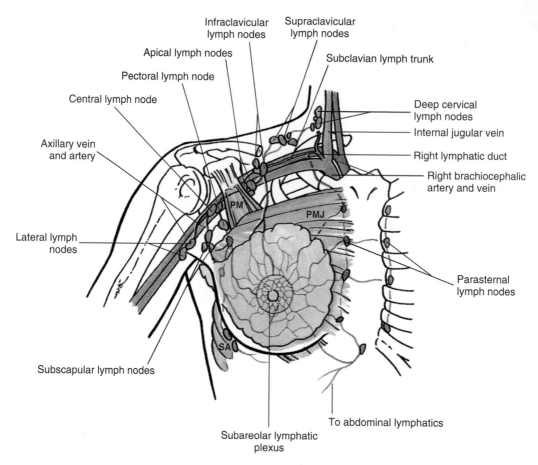

FIGURE 1.2 Lymphatic drainage of the right breast. Key Box: PM, pectoralis minor muscle; PMJ, pectoralis major muscle; SA, serratus anterior muscle.

D. Muscles, Nerves, and Vessels. Although we will not dissect the cutaneous nerves of the thoracic wall, please note their distribution in your atlas and appreciate that they are divided into lateral and anterior branches (Fig. 1.3). **A1 14:03-20:36/G1.20/C13/R202, 214/ N166, 179**

To better study the thoracic wall, we will first dissect two muscles of the upper limb, the pectoral muscles, and then proceed with the intercostal muscles.

1. Pectoralis major. Clean the pectoralis major muscle. This muscle consists of two heads: a large sternocostal head and a smaller clavicular head arising from the medial half of the clavicle. Identify the **deltopectoral triangle** and in this space the cephalic vein, which drains blood into the axillary vein. The pectoralis major muscle forms the anterior axillary fold (grip this muscle anterior to your own arm pit) and helps adduct the upper limb. **G1.2/C18/R195/ N174**

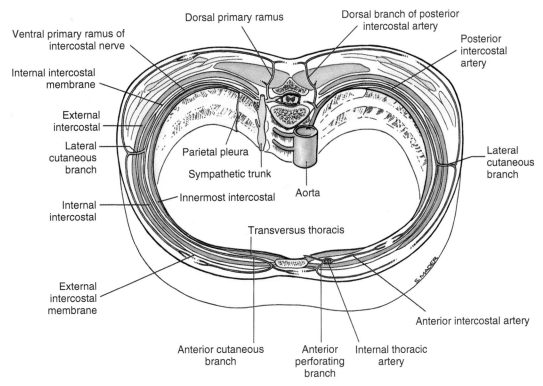

FIGURE 1.3 Transverse section of thorax showing contents of an intercostal space.

Cut and partially reflect the pectoralis major muscle close to its attachment to the clavicle. Look for the **lateral pectoral nerve** on its underside.   G6.13/C24/ R196/N174, 400

Next, relax the muscle by adducting the cadaver's arm. Separate the pectoralis muscle from the underlying clavipectoral fascia, which envelops the pectoralis minor muscle, by passing your fingers gently posterior to the sternal head. Now cut the sternal head of the muscle and begin reflecting the muscle laterally. Find the **medial pectoral nerve**, which pierces the pectoralis minor muscle before entering the pectoralis major. Cut the nerves and reflect the pectoralis major laterally toward the arm to expose the pectoralis minor (Fig. 1.4). **G6.13/C24/ R196/ N400**

2. **Pectoralis minor.** Identify and clean the **thoracoacromial artery** (a branch of the axillary artery) medial to the pectoralis minor muscle (it lies next to the lateral pectoral nerve). Next, detach the pectoralis minor from the costal cartilages of ribs 3, 4, and 5, and reflect it toward the shoulder where it inserts on the coracoid process of the scapula. Lateral to the pectoralis minor, locate the lateral thoracic artery, a branch of the axillary artery.   **G6.13/C20, 23/R196, 387/ N400**

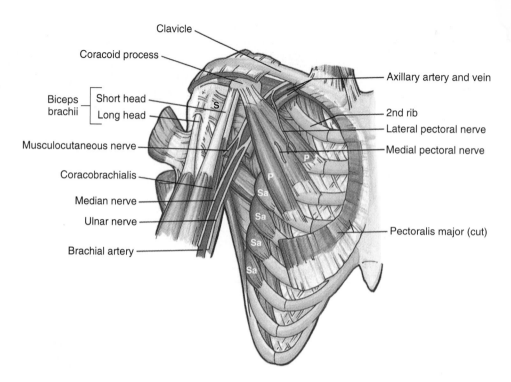

FIGURE 1.4 Anterior structures of thoracic wall. Key box: P, pectoralis minor; S, subscapularis; Sa, serratus anterior.

3. **Intercostal muscles.** Study the **external intercostal, internal intercostal,** and **innermost intercostals.** These muscles are supplied by the corresponding intercostal vessels and nerve, and fill the intercostal spaces. Study the distribution of these structures in your atlas and understand the function of the intercostal muscles in respiration (Table 1.1).   **A3 51:23–1:01:57/ G1.18, 1.20/C146-147/ R193-196/ N175, 177, 183**

In the fourth intercostal space (between ribs 4 and 5) at the anterior axillary line, identify each of the intercostal muscles by cutting through their layers and demonstrate an **intercostal nerve** and **vessels** (artery and vein) which lie just inferior to rib 4.

E. **Removal of Thoracic Wall.** To visualize the thoracic contents, we will first remove the breastplate.

Cut ribs 2 through 6 just *anterior* to the **serratus anterior muscle** on both sides (Fig. 1.5 Dashed lines mark the appropriate cuts) (Note: some instructors may prefer that you cut ribs 2 through 8, so please check with their preference). Cut only the ribs and not the underlying pleura, if possible. The parietal pleura often is adherent to the rib cage (occurs postmortem).

Muscles	Superior Attachment	Inferior Attachment	Innervation	Action[a]
TABLE 1.1 MUSCLES OF THORACIC WALL				
External intercostal	Inferior border of ribs	Superior border of ribs below	Intercostal n.	Elevate ribs
Internal intercostal	Inferior border of ribs	Superior border of ribs below	Intercostal n.	Depress ribs
Innermost intercostal	Inferior border of ribs	Superior border of ribs below	Intercostal n.	Probably elevate ribs
Transversus thoracis	Posterior surface of lower sternum	Internal surface of costal cartilages 2-6	Intercostal n.	Depress ribs
Subcostal	Internal surface of lower ribs near their angles	Superior borders of 2nd or 3rd ribs below	Intercostal n.	Elevate ribs
Levator costarum	Transverse processes of T7-T11	Subjacent ribs between tubercle and angle	Dorsal primary rami of C8-T11 nn.	Elevate ribs
Serratus posterior superior	Ligamentum nuchae, spinous processes of C7 to T3 vertebrae	Superior borders of 2nd to 4th ribs	Second to fifth intercostal nn.	Elevate ribs
Serratus posterior inferior	Spinous processes of T11 to L2 vertebrae	Inferior borders of 8th to 12th ribs near their angles	Ventral rami of ninth to twelfth thoracic spinal nn.	Depress ribs

[a]All intercostal muscles keep intercostal spaces rigid, thereby preventing them from bulging out during expiration and from being drawn in during inspiration. Role of individual intercostal muscles and accessory muscles of respiration in moving the ribs is difficult to interpret despite many electromyographic studies.

Using a saw, then cut horizontally through the manubrium just inferior to the first rib and cut the sternum at the sixth intercostal space (just above the xiphoid). Cut the internal thoracic vessels on both sides just superior and inferior to the breastplate. Gently elevate the inferior part of the sternum together with the attached portions of the severed ribs and cut any remaining soft tissue connections (Some instructors prefer to have the ribs cut as far posteriorly as possible to allow more access to the thoracic cavity. Please check with your instructor regarding their preference).

1. **Identify the following structures:** G1.16/ C162/R194/N176

    a. **Internal thoracic (mammary) artery** and **vein(s)**. These vessels anastomose with the posterior intercostal vessels laterally (Figs. 1.3 and 1.5).

    b. **Transversus thoracis muscle.**

    c. **Sternocostal joints** that allow gliding movements during respiration.

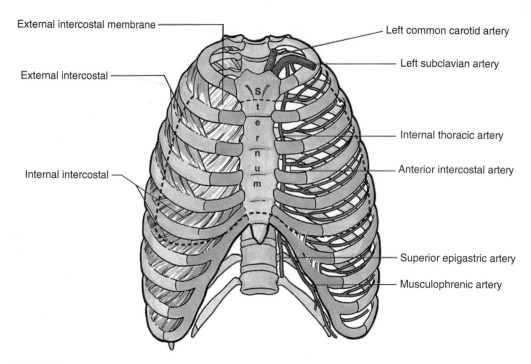

FIGURE 1.5 Anterior view of the thoracic wall. Dashed lines represent appropriate cuts. Note that on the left the internal thoracic artery is observed arising from the subclavian artery and runs about 1 cm lateral to the sternal border. On the right, observe the intercostal muscles.

II. PLEURAL CAVITIES AND THE LUNGS

Learning Objectives

- Understand the concept of pleural sac and potential space.
- Identify the features of each lung.
- Draw and describe the topographical surface projections of the lungs and parietal pleural reflections on the thoracic wall.
- Define the term "bronchopulmonary segment" and know its importance.

Key Concepts

- Potential space and pleural cavity
- Pleural reflections on the thoracic wall
- Bronchopulmonary segment

A. Introduction. The thoracic cavity contains two **pleural sacs** (containing the lungs) separated by the **mediastinum.** Each lung is covered with a smooth glistening membrane, the **visceral pleura.** At the root of the lung (where structures enter or leave the lung), the visceral pleura reflects off of the lung and forms the **parietal pleura,** which lines the walls of the pleural cavity (Fig. 1.6).

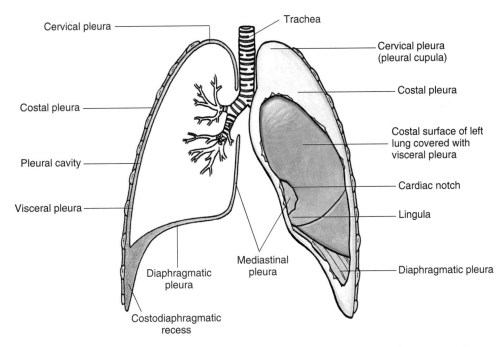

FIGURE 1.6 Pleurae and pleural cavities. On the left, costal pleura has been opened to view the lung covered with visceral pleura (pink). On the right, the bronchial tree is shown as it would appear entering the lung (lung absent on this side).

The parietal pleura is subdivided into the **costal** (lines rib cage), **mediastinal, diaphragmatic** (covers diaphragm), and **cervical (cupula)** pleura (extends into the neck).   **A3 46:01-51:22/ G1.21-1.22/ C164-165/ R232/N184-185**

Study the pleural reflections and the pleural recesses, where the lung does not intervene between two layers of parietal pleura. The two pleural cavities are two separate and closed potential spaces. Normally, they contain only a small amount of serous lubricating fluid that reduces friction during respiratory movements.

B. Pleural Cavities

If not already cut or damaged during removal of the rib cage, incise the parietal pleura longitudinally and explore the pleural cavity, pleural recesses (costodiaphragmatic and costomediastinal), lungs, and the root of the lung with your hand. You may encounter pleural adhesions (fusion of the visceral and parietal pleura due to inflammation), which can be broken down with your fingers. Over the diaphragmatic surfaces, palpate the underlying liver on the right and the spleen on the left.

C. Lungs

1. Identify the **phrenic nerves** (ventral rami of C3-5, "they keep the diaphragm alive") (innervates the abdominal diaphragm), which are *closely applied to the sides of the pericardial sac* and lie just anterior to the root of each lung. Since the diaphragm is skeletal muscle, the phrenic nerves are somatic nerves. *Save these nerves.*   G1.44/C190/R251/N182

Push the lung laterally to stretch the root of the lung and carefully transect the root structures (bronchi, pulmonary arteries, and veins) between the lung and mediastinum. Remove each lung.

2. **Lungs.** Each lung is divided into a **superior** and **inferior lobe** by an **oblique fissure.** The superior lobe of the right lung is further subdivided by a **horizontal fissure,** producing a small **middle lobe.** Note that the inferior lobe of each lung occupies the posterior part of the thoracic cavity while the superior lobe of the lung occupies the anterior part.   G1.30/ C172-173/ R233/ N186-187

   Observe the relative positions of the **bronchus, pulmonary artery,** and **pulmonary veins** at the **hilum** of the lung. Usually, the bronchus lies posterior, the artery superior, and the veins inferior. Identify the **pulmonary ligament,** where visceral pleura reflects off of the lung to become mediastinal pleura. Note numerous blackened **pulmonary lymph nodes** surrounding the hilum. Autonomic nerve fibers and small bronchial arteries also enter the lung at the hilum.   G1.28-1.29/C176-177/R233/ N187, 197

   Identify the **lobar (secondary) bronchi** in the left (superior and inferior) and right (superior, middle, and inferior) lung. G1.30-1.32/C181/R231/ N188-191

   On one lung, select a lobar bronchus, dissect deep into the lung tissue and identify **segmental (tertiary) bronchi** and branches of the pulmonary artery supplying a bronchopulmonary segment. A **bronchopulmonary segment** is that portion of the lung aerated by a tertiary bronchus and supplied by a single branch of the pulmonary artery. Each lung has 10 bronchopulmonary segments. Bronchioles and their accompanying arteries are intrasegmental while tributaries of the pulmonary veins lie at the periphery of a bronchopulmonary segment (intersegmental). Lymphatics follow both pathways.   G1.33/C180-181/R235/ N190-191

3. Identify the **descending aorta, intercostal arteries** arising from the **thoracic aorta,** and **intercostal nerves.**   G1.72/C189/ R262-263/N196, 219

4. Remove some costal parietal pleura adjacent to the bodies of several vertebrae and ribs, and identify the **sympathetic trunk, sympathetic chain ganglia,** and **rami communicantes** connecting the sympathetic trunk to the intercostal nerve (Fig. 1.7).   G1.70-1.71/C233/R262/ N228

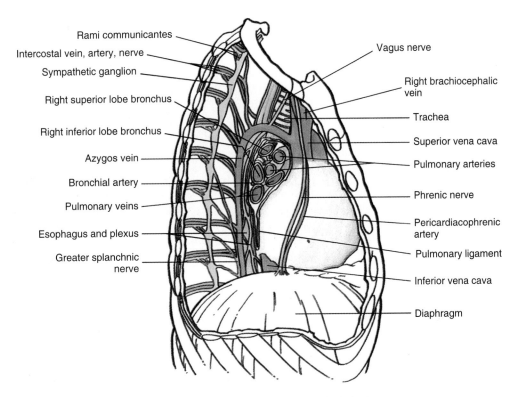

Rami communicantes
Intercostal vein, artery, nerve
Sympathetic ganglion
Right superior lobe bronchus
Right inferior lobe bronchus
Azygos vein
Bronchial artery
Pulmonary veins
Esophagus and plexus
Greater splanchnic nerve

Vagus nerve
Right brachiocephalic vein
Trachea
Superior vena cava
Pulmonary arteries
Phrenic nerve
Pericardiacophrenic artery
Pulmonary ligament
Inferior vena cava
Diaphragm

FIGURE 1.7   Right side of the mediastinum after removal of the lung.

5. Study the topographical projections of the parietal pleura and lungs on the thoracic wall (Fig. 1.8). This is important in "visualizing" these structures as they relate to the surface anatomy of the chest, especially as part of a physical examination.   **G1.25/C164-165/R232/N184-185**

FIGURE 1.8   Surface projection of pleurae and lungs.

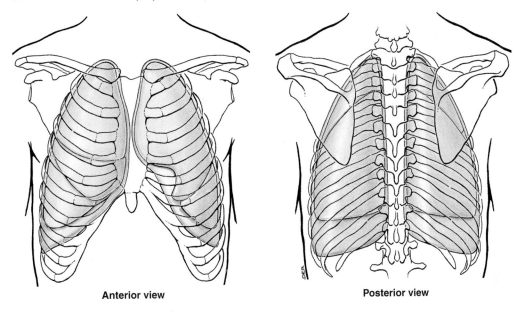

Anterior view   **Posterior view**

III. MEDIASTINUM AND HEART

Learning Objectives

- Understand the concept of "mediastinum" and define its boundaries.
- Know the concept of pericardial sac and potential space.
- Identify the features of the adult heart, both externally and internally.
- Learn the distribution of the coronary arteries.
- Describe the differences between fetal and adult circulation and identify circulatory changes that occur at or shortly after birth.
- Explain the concept of "referred pain" as it relates to myocardial ischemia.
- Define the term "anastomosis."
- Diagram the conduction system of the heart naming the key nodes and conduction bundles.

Key Concepts

- Mediastinum and its divisions
- Pericardial cavity as a potential space
- Anastomosis
- Referred pain from the heart
- Fetal circulation

A. **Introduction.** Depending on your laboratory schedule, this dissection may require one or two laboratory periods. Check with your instructor.

The central region between the two pleural sacs is called the mediastinum. From your textbook and atlas, note that the mediastinum is subdivided anatomically into four parts (clinicians often simplify this description and do not use the term superior mediastinum). Realize that in a living person the mediastinum is a mobile region accommodating changes due to movement and volume.

B. **Pericardium**

1. The middle mediastinum contains the heart, roots of the great vessels, and the ensheathing pericardium. From descriptions in your atlas and textbook, note the layers of the pericardium (a **fibrous pericardium,** and a **serous pericardium** composed of a parietal and visceral layer).   **G1.26, 1.44/C190/R250/N200-201**

2. With a forceps, pinch up a fold of pericardium and with scissors nick the fold and open the pericardial sac widely. Remove the entire anterior portion of the pericardium.

Define the **transverse pericardial sinus** (a space) by placing a finger between the superior vena cava and ascending aorta and pushing it posterior to the pulmonary artery. Push two fingers posterior to the heart's ventricles into the **oblique pericardial sinus.** This space is a serous-lined cul-de-sac bounded by the inferior vena cava and the four pulmonary veins.   G1.53/C194/R254-255/ N203

C. Heart and Great Vessels

1. Before removing the heart, study the following features:
G1.45/C190, 192/ R251/ N201-203

a. **Right atrium** and **right ventricle.**

b. Coronary arteries (to be dissected later) embedded in fat in the epicardium (visceral layer of serous pericardium).

c. **Superior vena cava** (SVC), **ascending aorta,** the **left vagus nerve** (10th cranial nerve) as it crosses the aortic arch, and the **recurrent laryngeal nerve** (innervates the larynx) and its relationship to the **ligamentum arteriosum.**

2. Remove the heart by cutting across the ascending aorta and pulmonary trunk at the level of the transverse pericardial sinus. Next cut the **inferior vena cava** (IVC) as inferior as possible within the pericardial sac. Cut the superior vena cava 1 cm above the right atrium and lift the heart anteriorly and superiorly by its apex and cut across the four pulmonary veins. Remove the heart.

3. Identify the **coronary** or **atrioventricular groove** (**sulcus**) separating the atria from the ventricles and the **interventricular grooves** which separate the ventricles. Identify the aorta and look inside its lumen to identify the **aortic valve.** Also, identify the **pulmonary trunk** and its valve, and the **superior vena cava.**   G1.45, 1.58/C199-200/ R236/ N202, 210

D. Cardiac Vessels. Carefully dissect through the epicardial fat to identify the cardiac vessels (Fig. 1.9). Use the "scissor technique" as you bluntly dissect in parallel along the artery.

1. Left coronary artery. Locate the opening of the left coronary artery just superior to the left aortic valve cusp and, beginning your dissection here, find the short **left coronary** and clean its two main branches, the **anterior interventricular** (**left anterior descending** or **LAD,** as it is called by most physicians) and the **circumflex branch.** Follow the circumflex branch around the left border of the heart.   G1.46, 1.58/C195-200/ R244/ N204-207

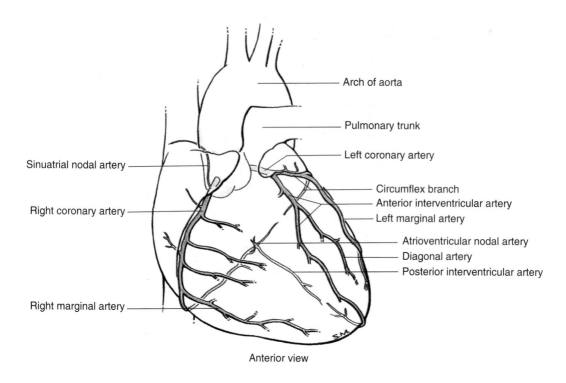

Arch of aorta

Pulmonary trunk

Left coronary artery

Sinuatrial nodal artery

Circumflex branch
Anterior interventricular artery
Left marginal artery

Right coronary artery

Atrioventricular nodal artery
Diagonal artery
Posterior interventricular artery

Right marginal artery

Anterior view

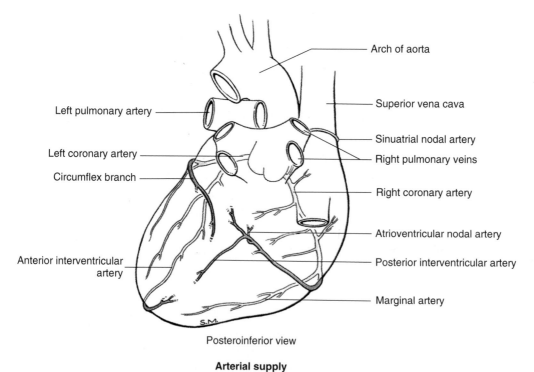

Arch of aorta

Left pulmonary artery

Superior vena cava

Sinuatrial nodal artery

Left coronary artery

Right pulmonary veins

Circumflex branch

Right coronary artery

Atrioventricular nodal artery

Anterior interventricular
artery

Posterior interventricular artery

Marginal artery

Posteroinferior view

Arterial supply

FIGURE 1.9   Arterial supply of the heart

2. **Right coronary artery.** Similarly, find the opening of the right coronary artery above the right aortic valve cusp and dissect the artery as it runs in the atrioventricular groove. It usually gives off a **marginal branch** and then ends on the posterior side of the heart as the **posterior interventricular branch** (called the **posterior descending** by most physicians). Try to find a small **right atrial branch,** which courses toward the superior vena cava and supplies the sinuatrial node (SA nodal branch). **G1.46/C195-200/ R244/ N204-207**

3. **Cardiac veins.** Identify the **great cardiac (anterior interventricular) vein** and **middle cardiac (posterior interventricular) vein,** which course with the LAD and posterior descending arteries, respectively. These veins drain into the **coronary sinus,** which then drains into the right atrium posteriorly. **G1.49/C197-198/R244/ N204-205**

E. **Interior of the Heart**

1. **Right atrium.** Open the right atrium by following the anterior incisions shown in Figure 1.10. Wash out the right atrium with water and identify the following features: **G1.55/C207-208/R240/ N208**

   a. **Pectinate muscles** (L. pecten, comb) rough muscular ridges on the anterior atrial wall and auricle (part of the original embryonic atrium).

   b. **Crista terminalis** , a vertical ridge separating anterior and posterior walls.

   c. **Fossa ovalis,** an oval depression in the interatrial wall.

   d. **Superior** and **inferior vena cava,** and opening for the **coronary sinus.**

   e. **Valve of the coronary sinus.**

   f. **Right atrioventricular (tricuspid) valve** and its **commissures** (site of union between valve leaflets).

   g. Although not visible grossly, the SA node lies in the junction of the crista terminalis and the SVC, while the AV node lies near the opening of the coronary sinus.

2. **Right ventricle.** Open the right venticle as shown in Figure 1.10. Pass a blunt instrument or your finger into the pulmonary trunk and determine the level of the pulmonary valve. Make your incision in the right ventricular wall just inferior to the pulmonary valve and extend the cut as shown in Figure 1.10. Turn the ventricular flap inferiorly to open the chamber widely. Carefully clean out blood clots from the

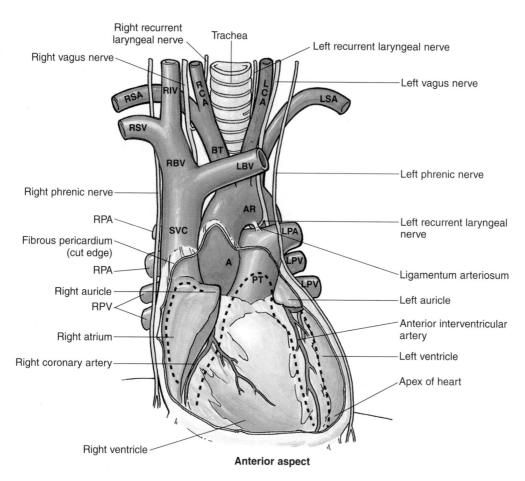

Anterior aspect

A	Aorta
AR	Arch of aorta
BT	Brachiocephalic
IVC	Inferior vena cava
LBV	Left brachiocephalic vein
LCA	Left common carotid artery
LPA	Left pulmonary artery
LPV	Left pulmonary vein
LSA	Left subclavian artery
PT	Pulmonary trunk
RBV	Right brachiocephalic vein
RCA	Right common carotid artery
RIV	Right internal jugular vein
RPA	Right pulmonary artery
RPV	Right pulmonary vein
RSA	Right subclavian artery
RSV	Right subclavian vein
SVC	Superior vena cava

FIGURE 1.10   Anterior heart. Dashed lines show incisions for opening the heart chambers.

chamber, rinse with cold water, and identify the following features: **G1.56/C207/ R240, 243/ N208**

a. **Tricuspid (right atrioventricular, RAV) valve** leaflets or cusps.

b. **Chordae tendineae.** Attached to the cusps and arranged like cords of a parachute.

c. **Papillary muscles.** The anterior papillary muscle is the largest and its chordae tendineae attach to the anterior and posterior valve cusps.

d. **Septomarginal trabecula (moderator band).** Stretches from the interventricular wall to the base of the anterior papillary muscle.

e. **Trabeculae carneae** (L. little cords of flesh). Rough muscular ridges.

f. Smooth-surfaced cone-shaped region called the **conus arteriosus** or **infundibulum** just inferior to the opening of the pulmonary trunk.

g. **Pulmonary semilunar valve.** Has three cusps: anterior, right and left.

3. **Left atrium.** Open the left atrium as shown in Figure 1.11 by making an inverted U-shaped incision through the posterior wall. Turn the flap inferiorly, clean and rinse the chamber, and identify the following features: **G1.60/C209/ R240/N209**

a. Openings for the **four pulmonary veins.**

b. **Left atrioventricular** or **mitral valve** and its **commissures.**

c. Pectinate muscle in the small left auricle.

4. **Left ventricle.** Open the left ventricle as shown in Figure 1.10. Start at the aorta and cut behind the pulmonary trunk and to the left of the LAD. The circumflex coronary branch will be cut. Extend the cut to the apex. Identify the following features: **G1.57/C209-210/R240/ N209, 212**

a. **Mitral (left atrioventricular, LAV) valve** and its **chordae tendineae.** It has an anterior and posterior cusp.

b. **Aortic semilunar valve,** and the **nodule,** a small fibrous thickening at the middle of the free margin of each valve cusp.

c. Openings of the two coronary arteries.

d. Place your fingers on either side of the **interventricular septum** and palpate the thick muscular and thinner membranous portions (only about the size of your finger nail). This can be the site of ventricular septal defects, the most common congenital cardiac defect.

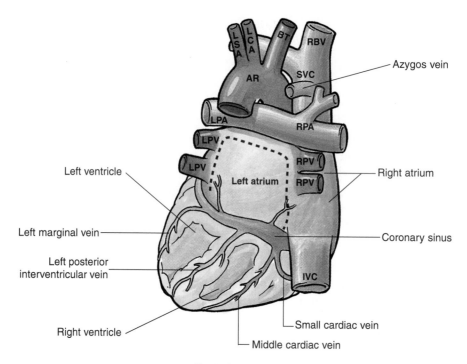

Posterior aspect

AR	Arch of aorta
BT	Brachiocephalic
IVC	Inferior vena cava
LBV	Left brachiocephalic vein
LCA	Left common carotid artery
LPA	Left pulmonary artery
LPV	Left pulmonary vein
LSA	Left subclavian artery
RBV	Right brachiocephalic vein
RPA	Right pulmonary artery
RPV	Right pulmonary vein
SVC	Superior vena cava

FIGURE 1.11   Dashed lines show incision in the left atrium.

e. Observe that the right ventricle is crescent-shaped in cross section because the interventricular septum is convex toward the right chamber. Also, the left ventricular wall is much thicker than the right chamber's wall due to the higher pressure.

5. In your textbook and atlas, review fetal and adult circulation through the heart and become familiar with circulatory changes that occur at birth or shortly thereafter. Also review the conduction system of the heart.   **G1.60/C215-218, 223-224/ R238, 243, 271/ N213, 217**

IV. POSTERIOR AND SUPERIOR MEDIASTINUM

Learning Objectives

- Describe the subdivisions of the mediastinum and their contents.
- Identify the great vessels in the superior mediastinum.
- List important structures associated with the sternal angle of Louis.
- Diagram the caval and azygos venous return to the heart.
- Describe the autonomic innervation to the thoracic viscera.
- List key collections of lymph nodes which drain the thorax.

Key Concepts

- Mediastinum
- Dual venous return via caval and azygos veins

A. Posterior Mediastinum

1. Remove any remaining pericardial sac and clean the esophagus. Note its important relationship to the left atrium of the heart and observe that the two vagal nerves "jump" onto the esophagus and form the esophageal plexus.   G1.37-1.38/C229, 237/R256/ N220

2. Identify again the **left vagus nerve, recurrent laryngeal nerve,** and **ligamentum arteriosum.** Find the left vagus in the interval between the **left common carotid** and **subclavian arteries.** Clean and follow the left vagus nerve as it passes over the aortic arch and posterior to the root of the lung. G1.38/C189/R261/ N219

   Pull the esophagus to the left and expose the **azygos vein** on the right. Find **intercostal veins** that drain into the azygos vein and study the azygos system of veins in your atlas. Note that the right vagus nerve courses along the trachea and then passes posterior to the root of the right lung.   G1.71-1.74/C233-234/ R261/ N218, 226

   Find the **thoracic lymphatic duct** in the space between the esophagus and descending thoracic aorta, or just anterior to the lower thoracic vertebrae in the midline.   G1.69, 1.71/C233-234/ R259/ N226-227

   Identify the **descending aorta** and its **intercostal branches.** Attempt to locate at least one of several bronchial arteries. G1.68/C231/ R263/ N196, 225

Remove some of the costal pleura lateral to the mid-thoracic vertebrae to reveal components of the sympathetic nerves (Fig. 1.7). Find the **sympathetic trunk** and look for the **greater splanchnic nerves,** which arise from T5-T9. The much smaller lesser splanchnic nerves (T10-T11) and least splanchnic nerves (T12) are difficult to find but note their course from your atlas. These splanchnic nerves pierce the diaphragm on their way to abdominal autonomic ganglia where they synapse (Fig. 1.12).
G1.70-1.71/C233/ R262-263/ N198, 228

B. Superior Mediastinum. Review the boundaries of the superior mediastinum (Fig. 1.13) and then with a bone pliers or saw, remove the inferior portion of the manubrium of the sternum to gain wider access to the area. Leave the first sternocostal joints intact. (Some instructors may prefer to remove the entire manubrium and anterior portions of the first rib).

FIGURE 1.12 Autonomic nerves in the posterior mediastinum. Orange, sympathetic; green, parasympathetic; blue, plexus.

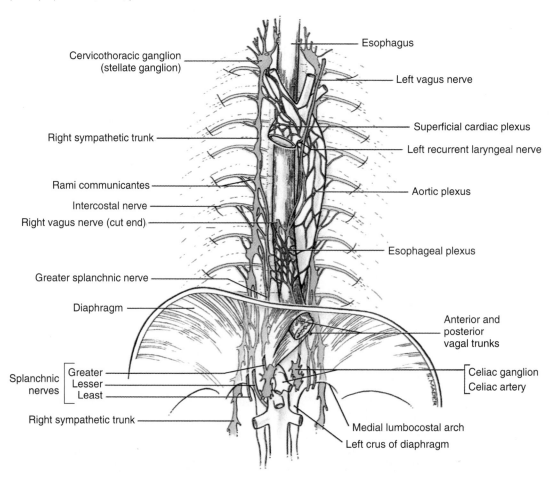

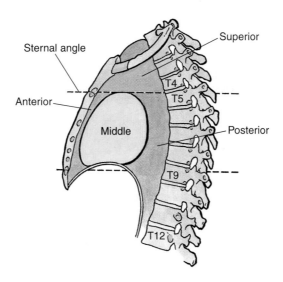

FIGURE 1.13 Subdivisions of the mediastinum.

1. Identify the **thymus gland** (atrophied due to age), which in adults is dispersed within a fatty mass just posterior to the manubrium.   **G1.62/C163/R248, 250/ N200**

   Remove the fatty thymus and clean the **right** and **left brachiocephalic veins** and **superior vena cava** (SVC). Locate where the **azygos vein** drains into the SVC.   **G1.37, 1.74/C194, 234/R259, 261/N201, 226**

2. Pull the great veins superiorly to expose the **arch of the aorta.** Identify its three great arteries: **brachiocephalic trunk, left common carotid artery,** and **left subclavian artery.** Between the aortic arch and tracheal bifurcation look for a number of fine nerve branches that are part of the cardiac autonomic plexus (Fig. 1.12).   **G1.45, 1.64/C237/ R253/ N198**

3. Find the two **phrenic nerves** (C3, 4, and 5 "keep the diaphragm alive"). The left phrenic crosses the aortic arch while the right phrenic descends lateral to the SVC. Both nerves pass anterior to the root of the lung and innervate the diaphragm.   **G1.38, 1.63/C190, 192/ R253/ N200, 214**

4. Identify the **bifurcation of the trachea** and look for **tracheobronchial lymph nodes.** Also note that the right main bronchus is more vertical, shorter, and wider than the left bronchus. With a scissors, cut the trachea transversely above the bifurcation and identify the midline **carina** (L. keel of a boat), a ridge inside the tracheal bifurcation.   **G1.64, 1.67/C187/ R257-258/ N190, 197, 227**

5. In your textbook and atlas, review the cross-sectional anatomy of the thorax paying special attention to the verti-

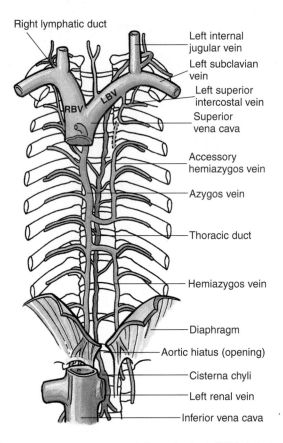

Right lymphatic duct

Left internal jugular vein

Left subclavian vein

Left superior intercostal vein

Superior vena cava

Accessory hemiazygos vein

Azygos vein

Thoracic duct

Hemiazygos vein

Diaphragm

Aortic hiatus (opening)

Cisterna chyli

Left renal vein

Inferior vena cava

RBV

LBV

FIGURE 1.14 Azygos venous system and thoracic duct. RBV, right brachiocephalic vein; LBV, left brachiocephalic vein.

cally running structures in the superior and posterior mediastinum. Also, review radiographs in lab or your textbooks and relate the images to the anatomy of the thorax. **G1.22, 1.39/C168/ R228, 268-269/ N180, 230**

Understand the importance of the dual venous return to the heart via the IVC and the azygos (azygos and hemiazygos) system of veins for blood coming from below the diaphragm or posterior thoracic wall (Fig. 1.14). **G1.74/C232-234/ R261/ N226**

2

ABDOMEN

I. ANTERIOR ABDOMINAL WALL

Learning Objectives

- Throughout Chapter 2, identify structures in bold print unless instructed to do otherwise.
- On a diagram, draw the regions and quadrants that divide the anterior abdominal wall.
- Identify the key anatomical features of the inguinal region and describe the differences between a direct and indirect inguinal hernia.
- Name the fascial layers contributing to the rectus sheath above and below the umbilicus (arcuate line).
- Know the functions of the abdominal wall musculature.

Key Concepts

- Significance of descriptive divisions of anterior abdominal wall
- Types of inguinal hernias
- Rectus sheath
- Superficial epigastric veins

A. Landmarks and Surface Anatomy

1. Palpate the following: A3 1:23:00-1:31:10/ G2.2/C282-284/ R183/ N231
   a. **Xiphisternal junction.**
   b. **Costal margin.** Upturned ends of cartilages 7-10.
   c. **Pubic symphysis.** Lower limit of anterior abdominal wall.
   d. **Pubic tubercle.**

e. **Anterior superior iliac spine.**

f. **Inguinal ligament.** Between the pubic tubercle and the anterior superior iliac spine.

For descriptive purposes, the anterior abdominal wall is divided into regions or quadrants. Study these descriptive subdivisions in your textbook and atlas. **C1/ R205/ N251**

Also, note key dermatomes of the anterior body wall: **T4** is the level of the nipple in males (variable in females), **T10** is the level of the umbilicus, and **L1** is the level of the pubic bone.

B. Muscles of the Anterior Wall—External Oblique

1. With the cadaver into the supine position, place a block beneath the lumbar region to stretch the anterior wall. First, make an incision from A to the B (Fig. 2.1). Then make transverse incisions from A to C, and from B to D in the proximal thigh. Reflect the skin flaps laterally all the way to the midaxillary line.

Note the superficial fascia (**Camper's fascia**), which contains variable amounts of fat and the membranous, deep layer of the superficial fascia (**Scarpa's fascia**), which lies superficial to the investing fascia of the abdominal muscles. Scarpa's fascia is composed of fibrous tissue and ends about 2 cm below the inguinal ligament where it attaches to the deep fascia of the thigh. This fascia also is continuous with the superficial perineal fascia (Colles' fascia) and dartos fascia of the penis and scrotum. Superficial veins (especially the **superficial epigastric veins**) lie in Camper's fascia. Note from your atlas the distribution of cutaneous nerves and realize that they are segmental continuations of the intercostal nerves from the thorax (dermatomes T7-L1). **A3 1:31:11-1:47:15/ G2.6/C247, 264/R197-198/ N232, 239-240**

FIGURE 2.1 Abdominal skin incisions.

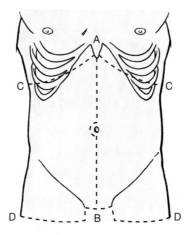

2. Clean the surface of the **external oblique.** Realize that the three flat muscles (external oblique, internal oblique and transversus) are continuations of the three intercostal muscles of the thorax, and are reinforced medially by the longitudinal strap-like rectus abdominis muscle. The rectus is enclosed in a fascial sheath composed of the aponeuroses of the three flat muscles. Note the **linea alba** in the midline where aponeurotic fibers (an aponeurosis is a broad tendon) of all three muscles from both sides interdigitate. **G2.6/C249/ R198/ N232**

C. Inguinal region. Although dissection instructions are provided for the male cadaver, those with female cadavers may follow the same procedure (but be sure to study from a male cadaver as well). This region is important because of the incidence of inguinal hernias, especially in males.

The inguinal canal is a 3-5 cm long passageway that lies just superior to the inguinal ligament along its medial half. It extends between the superficial ring (an opening in the external abdominal oblique) and the deep (internal) ring (an opening in the transversalis fascia). In the male, the ductus (vas) deferens traverses the canal. The canal passes through the arched aponeurotic and muscular layers of the anterior abdominal wall formed by the three flat muscles (Fig. 2.2).

1. Clean the aponeurosis of the **external oblique** in the inguinal region and identify the following features: **A3 1:47:16-1:54:09/ G2.8/C250-251/R205-208/ N232-234, 242-245, 351**

FIGURE 2.2 Inguinal region.

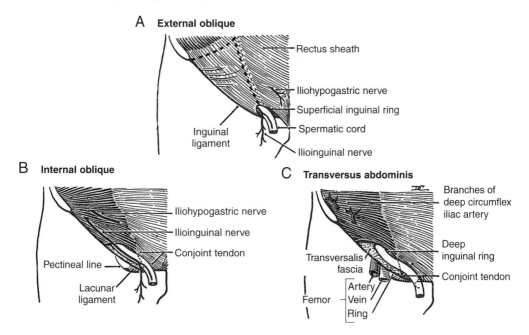

A **External oblique**
- Rectus sheath
- Iliohypogastric nerve
- Superficial inguinal ring
- Spermatic cord
- Ilioinguinal nerve
- Inguinal ligament

B **Internal oblique**
- Iliohypogastric nerve
- Ilioinguinal nerve
- Conjoint tendon
- Pectineal line
- Lacunar ligament

C **Transversus abdominis**
- Branches of deep circumflex iliac artery
- Deep inguinal ring
- Conjoint tendon
- Transversalis fascia
- Femor
- Artery
- Vein
- Ring

a. **Inguinal ligament** (Poupart's ligament). Inferior border of the aponeurosis of the external oblique rolled under on itself. It is attached laterally to the anterior superior iliac spine and medially to the pubic tubercle (Fig. 2.2A).

b. **Superficial inguinal ring.** Look for the spermatic cord as it leaves the ring to pass into the scrotum (in the female the round ligament of the uterus passes into the labia majora).

c. **Spermatic cord.**

d. **Lateral (inferior) crus** (leg) and **medial (superior) crus.** The spermatic cord lies on the inferior portion of the lateral crus. The medial crus is that portion of the aponeurosis that attaches to the pubic bone and crest medial to the pubic tubercle.

e. **Intercrural fibers.** Arch over the superficial ring and prevent the crura from spreading apart.

f. **Ilioinguinal nerve** (L1). Emerges from the superficial inguinal ring just lateral to the spermatic cord (Fig. 2.2B). This nerve is composed of sensory fibers from the external genitalia and upper medial thigh.

Further open the inguinal canal by cutting the external oblique aponeurosis as shown by the dashed line in Figure 2.2A. Identify the **conjoint tendon** (falx inguinalis), which is the fused internal abdominal oblique and transversus abdominis aponeuroses medial to the inguinal canal (Fig. 2.3). The **lacunar ligament** (Gimbernat's ligament) is the

FIGURE 2.3 Spermatic cord emerging from the superficial ring.

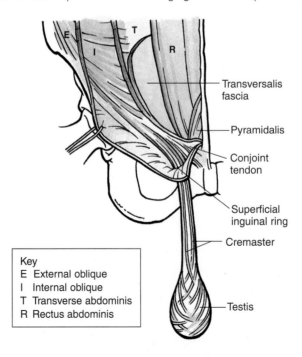

Key
E External oblique
I Internal oblique
T Transverse abdominis
R Rectus abdominis

crescent-shaped attachment of the fibers of the medial inguinal ligament to the pecten pubis (difficult to discern) (Fig 2.2B).   **G2.6/C269/R205-206/N242-244**

Consult your textbook regarding the clinical anatomy of indirect and direct inguinal hernias.

II. SCROTUM, SPERMATIC CORD, AND TESTIS

Learning Objectives

- Describe which layers of the anterior abdominal wall contribute to the coverings of the spermatic cord.
- List the contents of the spermatic cord.
- Identify the features of the testis and describe its blood supply and lymphatic drainage.
- List the contributions of the aponeuroses of the three flat abdominal muscles to the formation of the rectus sheath.

Key Concepts

- Spermatic cord contents
- Descent of the testis
- Rectus sheath
- Superficial venous connections for venous return to the heart

A. Introduction. The scrotum, spermatic cord, and testis are closely related to the inguinal canal and will be dissected at this time. If you have a female cadaver, join a table with a male cadaver for this dissection exercise.

The **scrotum** is a cutaneous outpouching of the anterior abdominal wall. The superficial fascia of the scrotum is void of fat but contains smooth muscle (**dartos**), which controls the surface area of the scrotum. The scrotum is divided into right and left halves and, during development, the testicles descend into their respective scrotal halves and become anchored to it inferiorly by the gubernaculum.

B. Scrotum and Spermatic Cord. Make an incision down the lateral scrotum through the skin, dartos, and superficial fascia. Free the testis and spermatic cord from the surrounding connective tissue and remove the testis by cutting the band of tissue that anchors the testis to the scrotum (**gubernaculum testis**).

1. Carefully incise the coverings of the spermatic cord longitudinally and identify (Fig. 2.4):   **G2.17/C271-274/ R206, 321/ N361**

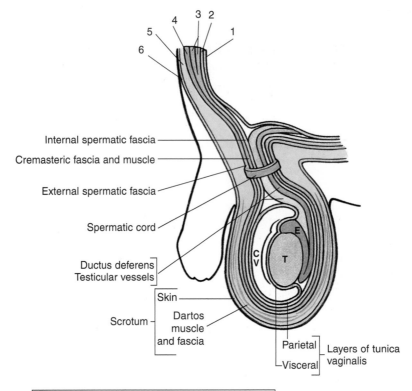

FIGURE 2.4   Coverings of spermatic cord and testis.

Internal spermatic fascia
Cremasteric fascia and muscle
External spermatic fascia
Spermatic cord
Ductus deferens
Testicular vessels
Skin
Scrotum
Dartos muscle and fascia
Parietal ⎤ Layers of tunica
Visceral ⎦ vaginalis

Key	
1	Peritoneum
2	Transversalis fascia
3	Transversus abdominis, internal oblique
4	External oblique
5	Subcutaneous fat
6	Skin
E	Epididymis
CV	Cavity of tunica vaginalis
T	Testis

a. **External spermatic fascia.** Formed from the aponeurosis of the external abdominal oblique muscle.

b. **Cremasteric fascia.** Contains cremaster muscle fibers derived from the internal abdominal oblique muscle.

c. **Internal spermatic fascia.** Derived from the transversalis fascia.

d. **Ductus (vas) deferens,** and its small artery. The vas is identifiable by palpation as a firm tubular structure.

e. **Testicular artery.**

f. **Pampiniform plexus of veins.** Cools the arterial blood of the testicular artery via a counter-current mechanism.

C. Testis

1. Note the **tunica vaginalis testis,** a closed serous sac (potential space) of peritoneal origin. Identify the **tunica albuginea,** the dense white connective tissue capsule of the testis, and the **head, body,** and **tail of the epididymis,** and **efferent ductules.** The epididymis is attached to the superior and posterolateral surface of the testis (Fig. 2.5). Slice the testis longitudinally from its superior to inferior pole and note the thin strands of the **seminiferous tubules.**   G2.17/C274-275/ R321/ N361-362

D. Muscles of the Anterior Wall Continued (Table 2.1). Now return to the abdominal wall and finish the dissection of the remaining muscles.

1. **Internal oblique.** Identify this muscle in the region where you cut through the external oblique (Figs. 2.2A and 2.3). In the inguinal region, the tendinous fibers of the internal oblique join the tendinous fibers of the transversus abdominis to form the conjoint tendon.   G2.11-2.12/C253, 270/ R200, 206/ N243

2. **Transversus abdominis.** Incise the internal oblique along the anterior axillary line to demonstrate the fibers of the underly-

FIGURE 2.5   Dissected view of the testis, epididymis and distal spermatic cord.

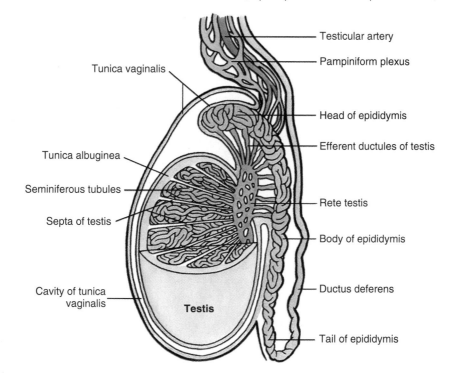

Testicular artery

Pampiniform plexus

Tunica vaginalis

Head of epididymis

Efferent ductules of testis

Tunica albuginea

Seminiferous tubules

Rete testis

Septa of testis

Body of epididymis

Cavity of tunica vaginalis

Ductus deferens

Testis

Tail of epididymis

TABLE 2.1
PRINCIPAL MUSCLES OF ANTEROLATERAL ABDOMINAL WALL

Muscles	Origin	Insertion	Innervation	Action(s)
External oblique	External surfaces of 5th to 12th ribs	Linea alba, pubic tubercle, and anterior half of iliac crest	Inferior six thoracic nn. and subcostal n.	Compress and support abdominal viscera; flex and rotate trunk
Internal oblique	Thoracolumbar fascia, anterior two-thirds of iliac crest, and lateral half of inguinal ligament	Inferior borders of 10th-12th ribs, linea alba, and pubis via conjoint tendon	Ventral rami of inferior six thoracic and first lumbar nn.	
Transversus abdominis	Internal surfaces of seventh to twelfth costal cartilages, thoracolumbar fascia, iliac crest, and lateral third of inguinal ligament	Linea alba with aponeurosis of internal oblique, pubic crest, and pecten pubis via conjoint tendon		Compresses and supports abdominal viscera
Rectus abdominis	Pubic symphysis and pubic crest	Xiphoid process and fifth to seventh costal cartilages	Ventral rami of inferior six thoracic nn.	Flexes trunk and compresses abdominal viscera

ing **transversus abdominis muscle.** These two muscles may be fused.   **G2.11/C255/R200-204/N243**

Beneath the transversus abdominis note the underlying **transversalis fascia,** a layer of fascia that lines the entire abdominal wall.

Retract the spermatic cord laterally and observe the **inferior epigastric vessels** shining through the transversalis fascia in the inguinal region. These arteries arise from the external iliac arteries and are landmarks for identifying direct and indirect inguinal hernias.   **G2.12/C266, 270/R206/ N243, 245**

3. **Rectus abdominis.** Open the anterior wall of the rectus sheath vertically to display its contents as shown in Figure 2.6. First, open the anterior lamina of the rectus sheath on the right side along the dashed lines between points A and B. Note that the rectus sheath is attached to the **rectus abdominis** at three **tendinous insertions (inscriptions).** Free these connections with your hand or scalpel. **G2.6/C253/ R201-204/ N233**

Note that anterior branches of six spinal nerves (T7-T12) supply the muscle by entering its lateral border. Just below the umbilicus, divide the anterior sheath and rectus muscle transversely from point C to D and reflect the two halves superi-

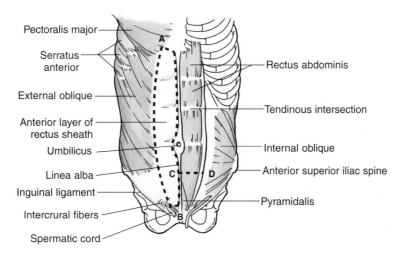

FIGURE 2.6   Dashed lines indicate cuts for opening the rectus sheath. On the left, the anterior lamina of the sheath has been removed to show the rectus abdominis muscle.

orly and inferiorly. On the deep surface, identify the **inferior and superior epigastric vessels** (**superior** vessels lie above the navel and **inferior** vessels below). The superior epigastrics connect with the internal thoracic vessels superiorly. On the posterior wall of the rectus sheath identify the **arcuate line,** below which the sheath is only composed of transversalis fascia posteriorly (Fig. 2.7).

FIGURE 2.7   Rectus sheath. Transverse sections of anterolateral abdominal wall superior to umbilicus (upper) and inferior to umbilicus (lower).

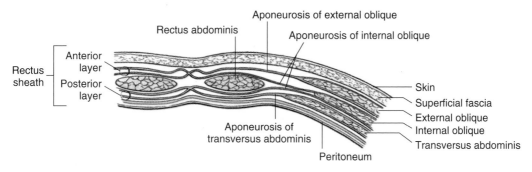

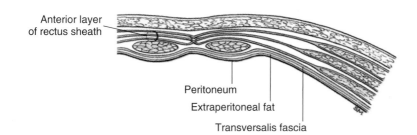

III. PERITONEUM AND PERITONEAL CAVITY

Learning Objectives

- Define what is meant by the peritoneal cavity, learn its subdivisions, and realize that it is a potential space.
- Identify the major organs of the abdominopelvic cavity and note surface features that are descriptive for localizing specific segments of the GI tract.
- Define the terms "omental bursa" and "mesentery."
- Understand the concept of retroperitoneal and distinguish between organs that are "secondarily" retroperitoneal from organs that are truly retroperitoneal.

Key Concepts

- Peritoneal cavity and greater and lesser sacs
- Mesenteries and peritoneal ligaments
- Retroperitoneal

A. Orientation. Understand that the **peritoneum** is a thin, translucent, serous membrane that lines the walls of the abdominal cavity, where it is known as **parietal peritoneum.** Organs that invaginate the peritoneal sac are invested with **visceral peritoneum.** The peritoneal cavity is a potential space containing only a small amount of lubricating serous fluid and everywhere in the cavity peritoneum is in contact with peritoneum. **Mesenteries** are two layers of peritoneum that "sling" the intestine from the posterior abdominal wall. Vessels, nerves, and lymphatics travel in mesenteries. Read about these concepts in your textbook as they are important to understand.

First, cut through the abdominal wall vertically along its length from the xiphoid process to the symphysis pubis keeping your incision about 1 cm to the left of the linea alba. Insert your finger through the incision and lift up the posterior lamina of the rectus sheath to avoid cutting the underlying abdominal viscera. Cut around the umbilicus on the left side. Then make horizontal incisions from the xiphoid process along the costal margin to rib 10 and cut inferiorly to the iliac crest. Fold the anterior abdominal wall inferiorly. Note on the right flap the attachment of the **falciform ligament** and the **ligamentum teres** (obliterated umbilical vein) in its free margin. These will need to be severed when the right abdominal flap is cut and reflected.

Initially, examine the peritoneal (inside) aspect of the umbilical region by identifying obliterated remains of fetal structures that radiate from the umbilical region. In the midline ascending from the urinary bladder is the obliterated allantoic duct (urachus) which forms the **median umbilical ligament.** The obliterated umbilical artery or **medial umbilical ligament** (one on each side of the midline) also ascends from the lower abdominopelvic cavity and can be observed on the inside aspect of the anterior abdominal wall.   G2.18/C262/R275/N236

B. Gastrointestinal Tract. Clean the surfaces of the viscera with a damp sponge and then inspect, but do not dissect, the GI tract from its beginning to its end and note the following features (Figs. 2.8 and 2.9):

1. **Stomach and Liver.** The stomach varies in shape and is attached to the liver by the **lesser omentum.** Identify the **fundus, body, antrum,** and **pylorus** of the stomach. The lesser omentum is divided into the **hepatogastric** and **hepatoduodenal ligaments.** Note the **greater omentum** (Fig. 2.8). It is attached to the **greater curvature of the stomach** and spreads apron-like over the transverse colon and covers portions of the small intestines. It varies in size and extent, in the living is mobile, and can wall off inflammed areas of the bowel to protect unaffected viscera.

   The **liver** is divided into **right** and **left lobes** by the **falciform ligament.** The **gallbladder** is attached to the inferior surface of the liver (Fig 2.8).   G2.18/C286/R273, 276/N252, 258

FIGURE 2.8   Anterior abdominal wall cut away to show undisturbed contents.

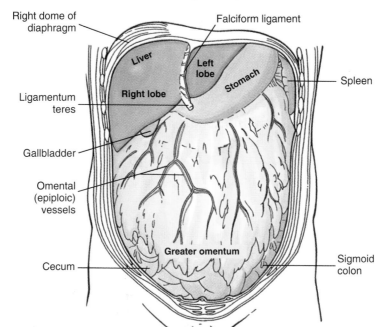

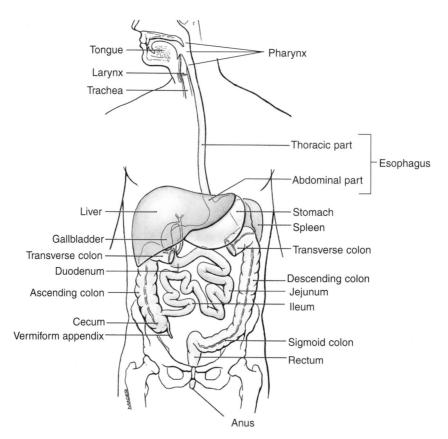

FIGURE 2.9 Digestive system.

2. **Small intestine.** The **duodenum** is largely secondarily retroperitoneal (has no mesentery) and inaccessible at this time. However, the proximal duodenum is mobile and is attached to the liver by the **hepatoduodenal ligament.** The remainder of the duodenum is C-shaped and molded around the head of the pancreas (also secondarily retroperitoneal). The terminal portion of the duodenum can be seen at the **duodenojejunal junction.** Pull the mobile small intestine to the right side and follow the proximal jejunum to see where it joins the duodenum.

The mesenteric small intestine consists of the proximal two-fifths called the **jejunum** and the distal three-fifths called the **ileum.** Find the root of the mesentery running diagonally across the posterior abdominal wall from upper left to lower right. The total length of the small intestine (about 20 feet) is accommodated in this small area. **G2.1-2.2/C287/R286-287/ N252-254**

3. **Large intestine.** Identify the **cecum** (L. caecus, blind) which extends inferiorly beyond the ileocecal junction. The vermiform appendix (L. vermis, worm; forma, shape; appendere, to

hang on) opens into the cecum and is suspended by a mesentery, the **mesoappendix** (most commonly found in a retrocecal position).   **G2.2, 2.62/C287/ R287-289/N254, 264**

The **ascending colon** (Gr. kolon, large intestine; hollow) is secondarily retroperitoneal. Identify the right colic or **hepatic flexure.**

The **transverse colon** runs from the hepatic to the **splenic** (left colic) **flexure.** Its mesentery is the **transverse mesocolon.** The left colic flexure is attached to the diaphragm by the **phrenicocolic ligament,** which acts like a shelf to support the spleen. The left colic flexure is more superior and posterior than the right colic flexure. Between these two flexures, the transverse colon is freely mobile.

The **descending colon** is secondarily retroperitoneal and often smaller in diameter than the ascending colon.

The **sigmoid colon** has a mesentery, the **sigmoid mesocolon,** and is very mobile.

The **rectum** (L. rectus, straight) is only partially covered by peritoneum and will be studied later.

On the colon's surface identify the outer longitudinal muscular bands known as **teniae coli** (three narrow bands), the sacculations called **haustra,** and the **appendices epiploicae** or small "bags of fat" suspended from the colon.   **G2.64, 2.68/C340/ R287/ N267**

C. Omental Bursa and Peritoneal Reflections

1. Place your finger in the **omental foramen** (epiploic foramen of Winslow). This foramen connects the omental bursa (lesser sac) (Fig. 2.10) with the greater sac or peritoneal cavity proper.   **G2.19/C299/ R291/ N271**

2. Cut the hepatogastric portion of the lesser omentum between the stomach and liver *but leave the hepatoduodenal ligament undisturbed.* Explore the lesser sac and realize that your hand lies between the stomach anteriorly and the retroperitoneally positioned pancreas posteriorly.   **G2.19-2.22/C291,303/R292/ N256**

3. Digitally examine the peritoneal attachments of the spleen by palpating the **splenorenal (lienorenal)** and **gastrosplenic ligaments,** which suspend the spleen between the kidney and stomach.   **G2.21/C303/R298/ N256-257**

4. Examine the peritoneal attachments of the liver: To gain better access, cut the right costal cartilages 6 and 7 near the xiphisternal junction and partially incise the diaphragm. The **falciform ligament, coronary ligaments,** and on the right side the **hepa-**

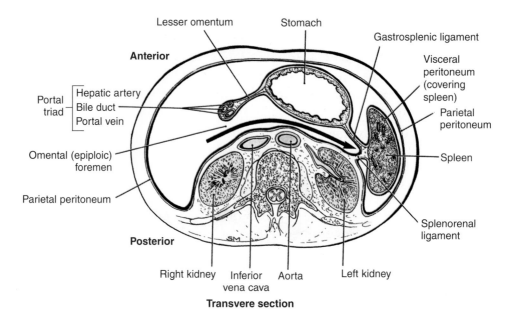

Lesser omentum   Stomach
                                              Gastrosplenic ligament
Anterior
                                              Visceral
                                              peritoneum
Portal ┌ Hepatic artery   (covering
triad │ Bile duct   spleen)
       └ Portal vein   Parietal
                                              peritoneum
Omental (epiploic)
foremen   Spleen

Parietal peritoneum

                                              Splenorenal
                                              ligament
Posterior

Right kidney   Inferior   Aorta   Left kidney
               vena cava
Transvere section

FIGURE 2.10   Transverse section seen from below. The arrow is in the lesser sac.

torenal ligament may now be examined. Inferior to the hepatorenal ligament is a space called the hepatorenal pouch or **recess of Morison,** bounded by the liver, right kidney, colon, and duodenum. This is the lowest point of the peritoneal cavity in a supine patient.   G2.18/C301/R280, 298/ N257, 270

IV. BILE PASSAGES, CELIAC TRUNK, AND PORTAL VEIN

Learning Objectives

- Describe, and be able to label on a diagram, the blood supply to the foregut derivatives of the GI system.
- Be able to trace bile from the liver to the gallbladder and from the gallbladder to the duodenum naming each duct traversed.
- Describe the portal venous system and important portal-caval or porta-systemic anastomoses.
- Compare and contrast the features of the portal and caval venous systems.
- Identify on the anterior abdominal wall the surface projections of the liver, stomach, pancreas, and spleen.

Key Concepts

- Bile duct system
- Portal venous system
- Portal-caval anastomoses

A. Introduction. This dissection will require time and patience. This region of the abdomen is critically important, both for the internist as well as the surgeon. We will dissect important structures in the hepatoduodenal ligament, find the branches of the celiac artery (supplies embryonic foregut derivatives in the abdomen), and trace the extrahepatic biliary system.

B. Common Bile Duct

1. Insert your finger in the epiploic foramen and realize that anterior to your finger lies the hepatoduodenal ligament and its contents: bile passages, hepatic artery, portal vein, autonomic nerves and lymphatics. To aid your dissection, place a paper towel in the epiploic foramen and then dissect the contents of the hepatoduodenal ligament. First, dissect the **common bile duct** on the right; it is thin-walled, usually stained green (a postmortem change), about the diameter of a pencil, and often collapsed. Dissect toward the gallbladder and identify the **cystic duct,** and the **right** and **left hepatic ducts.** (Variations to the normal pattern are common in this region and surgeons must always be aware of this). As you dissect, note the rich autonomic fiber plexus in the ligament and the presence of hepatic lymph nodes.   G2.28, 2.39/C292/ R294-296/ N276, 282

C. Hepatic Artery

1. To the left of the common bile duct, find and clean the **hepatic artery proper,** which arises from the **common hepatic artery,** a branch of the **celiac trunk** (Fig. 2.11). Follow the common

FIGURE 2.11 Hepatoduodenal ligament dissection.

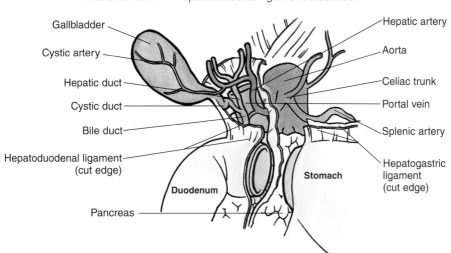

hepatic artery to the celiac trunk. Identify the **gastroduodenal artery** and **right gastric artery.** Finally, dissect the **left** and **right hepatic arteries** to the right and left lobes of the liver. In about 15% of cadavers, the right hepatic artery may arise from the superior mesenteric artery.   G2.98/C292/ R295-296/ N282-284

D. Portal Vein

1. Review the portal venous system in your textbook and atlas. It carries venous blood from the abdominal GI tract, spleen, and pancreas to the liver for metabolic processing. The portal vein is formed by the union of the splenic vein and superior mesenteric vein (Fig. 2.12). Dissect the proximal portion of the **splenic vein** as it lies inferior to the **splenic artery** and follow the vein to the **portal vein** (do not dissect the full extent of the splenic vein yet). Fold the stomach upward to gain access to

FIGURE 2.12   Portal venous system.

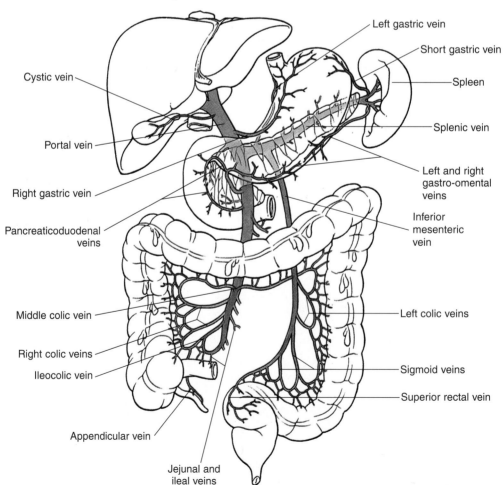

the vein. Sever the greater omentum from the stomach below the gastroepiploic vessels between its attachment to the greater curvature of the stomach and the transverse colon. Appreciate that you now have wide access to the **omental bursa** (space posterior to the stomach, see Fig. 2.10). Identify the **superior mesenteric vein.** Find where the **inferior mesenteric vein** drains into the portal system. This vein may join the splenic, superior mesenteric or join the other two at their junction where they form the portal vein. Look for **gastric veins** that carry blood from the esophagus and lesser curvature of the stomach to the portal vein.   G2.49/C298/ R282-283/ N293

E. Splenic Artery and Left Gastric Artery

1. Dissect the other two branches of the celiac trunk. First, dissect the large proximal portion of the **splenic artery** coursing from the celiac trunk along the superior border of the **pancreas** (Fig. 2.13). Follow the artery for 2 or 3 cm along the superior border of the pancreas. Then, clean the **left gastric artery** to the **lesser curvature of the stomach.** Note any lymph nodes lying along the splenic artery (pancreaticosplenic

FIGURE 2.13 Deep structures of the upper abdomen. Portion of the stomach removed.

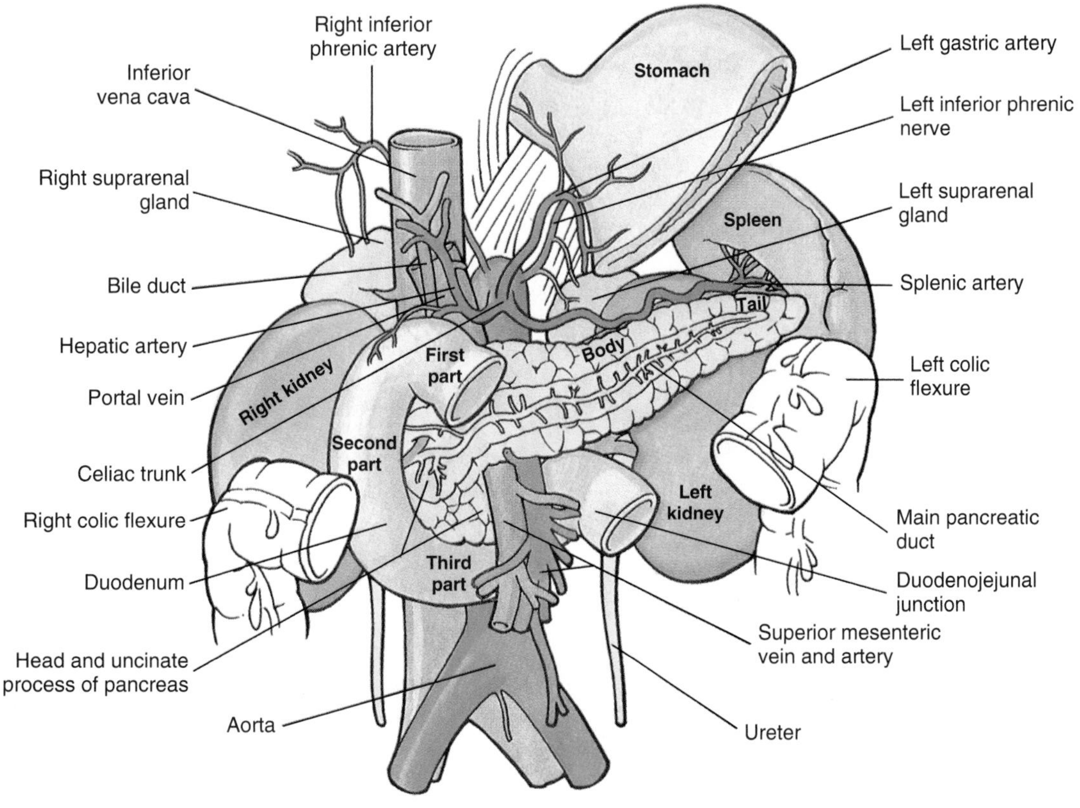

nodes). These nodes drain lymph into the para-aortic (lumbar) nodes adjacent to the aorta and celiac trunk.   **G2.34/C292-293/ R297/ N282-284**

2. Along the greater curvature, dissect the **right** and **left gastroepiploic arteries.** Pull the stomach superiorly to expose the omental bursa again and now complete the dissection of the splenic artery (Fig. 2.13). Pull the spleen anteriorly to facilitate exposure and trace the splenic artery to the **hilum of the spleen.** Note that it gives arterial branches to the body and tail of the pancreas and **short gastric branches** to the fundus of the stomach.   **G2.29/C292/ R294/N282-283**

F. Gallbladder and Ducts

1. Dissect the **common bile duct** through the **head of the pancreas** to the point where it enters the duodenum, called the **major duodenal papilla** (of Vater). Find the **main pancreatic duct** where it joins the common bile duct and note the thickened wall of smooth muscle, which comprises the **sphincter of Oddi** (Fig. 2.14). Remove tissue along the main pancreatic duct several centimeters into the substance of the pancreas.
G2.55, 2.57/C312-317/ R278, 302/ N276, 279, 284

FIGURE 2.14   Extrahepatic bile passages.

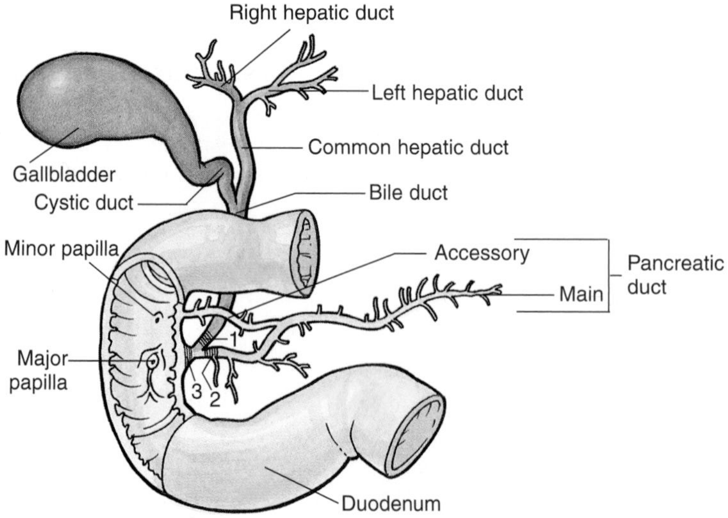

Key	
1	Choledochal sphincter
2	Pancreatic duct sphincter
3	Hepatopancreatic sphincter in wall of hepatopancreatic ampulla

2. On its anterior side, open the second part of the duodenum opposite the bile duct and find the **major duodenal papilla** and a hoodlike plica of tissue covering it (often difficult to see in the cadaver). If your cadaver has an accessory pancreatic duct, its opening will lie about 2 cm anterosuperior to the major papilla (also very difficult to find).

3. Return to the gallbladder, cut it open, and determine if gallstones are present. Observe the honeycombed mucosa typical of the gallbladder (not obvious if distended).

V. SUPERIOR AND INFERIOR MESENTERIC VESSELS

Learning Objectives

- Describe, and be able to label on a diagram, the blood supply to the midgut and hindgut derivatives of the GI tract.
- Describe the lymphatic drainage of the abdominal viscera.
- List points of anastomoses between branches of the SMA and IMA.

Key Concepts

- Pattern of blood supply and innervation to the foregut, midgut, and hindgut embryonic derivatives
- Lymphatic drainage to para-aortic (lumbar) nodes

A. Introduction. The celiac trunk (Fig. 2.13), superior mesenteric artery (SMA), and inferior mesenteric artery (IMA) (Fig. 2.15) are the three unpaired branches of the abdominal aorta and supply blood respectively to foregut, midgut, and hindgut embryonic derivatives of the GI tract and its associated organs.

B. Superior Mesenteric Artery

1. Reflect the **tail** and **body of the pancreas** to the right and find and clean the initial portion of the **SMA.** The SMA is as large as the celiac trunk and often lies 1 cm inferiorly. Follow the SMA until it crosses the third portion of the duodenum (2.13). The **superior mesenteric vein** lies just to the right of the SMA and drains the same area supplied by the artery. Note the plexus of autonomic nerve fibers around the SMA. **G2.68/C329/ R283-284/N285-286, 300**

2. Stretch the mesentery and palpate the SMA just to the right of the duodenojejunal junction. Using blunt dissection, clean the

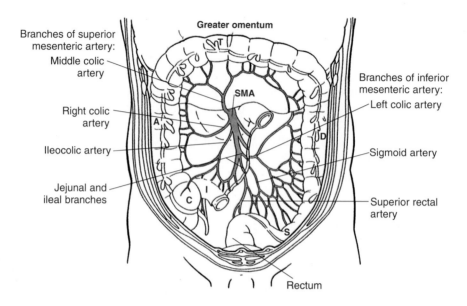

FIGURE 2.15 Superior and inferior mesenteric arteries.

following vessels in the mesentery: **G2.68-2.69/C329-330/R283/ N286-287**

a. **Intestinal arteries.** An array of 15-18 arteries to the jejunum and ileum. Note the vasa recta to the bowel margin.

b. **Ileocolic artery.** Supplying the cecum and appendix.

c. **Right colic artery.** Arising from either the SMA or ileocolic artery to supply the ascending colon.

d. **Middle colic artery.** Supplying the proximal two-thirds of the transverse colon.

e. Tributaries of the **superior mesenteric vein** correspond to the branches of the SMA.

C. Inferior Mesenteric Artery

1. The **IMA** arises from the aorta about 3 cm superior to the aortic bifurcation and is much smaller than the SMA. Find the

left colic artery, which supplies the distal transverse colon and descending colon, and then trace it back to the main trunk of the IMA. Also dissect the **sigmoidal arteries,** which form several arches to the sigmoid colon. Finally, find the **superior rectal artery,** which supplies the proximal portion of the rectum. Note the presence of the **marginal artery** (of Drummond), which provides anastomotic loops between branches of the SMA and IMA along the margin of the colon. **G2.72/C331/R285/ N287**

2. Tributaries of the **inferior mesenteric vein** correspond to the branches of the IMA. Lymph nodes and vessels follow the superior and inferior mesenteric veins and their branches, with the lymph ultimately collecting in para-aortic (lumbar) nodes along the midline. Look for representative lymph nodes in the mesentery of the small intestine and along the abdominal aorta (Fig. 2.16).   **G2.53, 2.98/C331/R282-283/N292-293, 296-297**

FIGURE 2.16 Lymphatic drainage of colon.

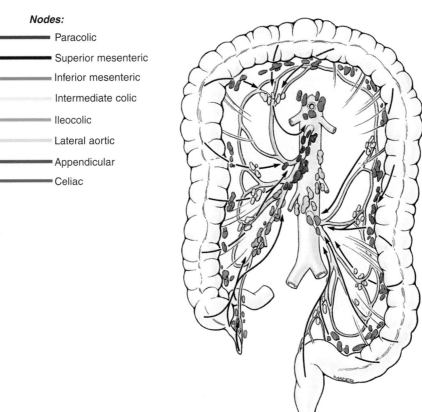

Nodes:

━━━ Paracolic

━━━ Superior mesenteric

━━━ Inferior mesenteric

━━━ Intermediate colic

━━━ Ileocolic

━━━ Lateral aortic

━━━ Appendicular

━━━ Celiac

VI. REMOVAL OF THE GI TRACT

Learning Objectives

- Identify the key features of the stomach, intestines, liver, and spleen.
- List the structures passing into or out of the porta hepatis.
- Demonstrate how the fetal circulation largely bypasses the liver using the ductus venosus.
- Describe the innervation of the foregut and midgut derivatives and relate this pattern of innervation to the fetal blood supply of these gut regions.
- Name the important abdominal wall surface relationships of the stomach, spleen, liver, and intestines.

Key Concepts

- Diagnostic features of the bowel
- Porta hepatis
- Bare area of the liver

A. Introduction. In this dissection, the abdominal GI tract will be removed *en bloc* while its three unpaired organs (liver, gallbladder, and spleen) will remain *in situ*. Later in the dissection, selected tables may remove the liver for further study. Some instructors may wish to remove the GI tract with the three unpaired organs while still others may prefer to keep some abdominal cavities undisturbed for future review by the students. Please check with your instructor regarding the procedure you should follow.

B. Removal of the GI Tract (Fig. 2.17). Key steps to be followed in this procedure are numbered below and in Figure 2.17 for orientation. Dashed lines show where cuts are made. Remove the GI tract in this sequence.

1. Begin by tying two strings about 2.5 cm apart *tightly* around the proximal rectum. Cut the rectum between the two strings.

2. Cut the IMA close to the abdominal aorta, leaving only a short stump for future reference.

3. Cut through the mesentery of the sigmoid colon on its *lateral* side so as not to damage the inferior mesenteric vessels.

4. Detach the descending colon with your fingers from the posterior abdominal wall. Cut through the phrenocolic ligament.

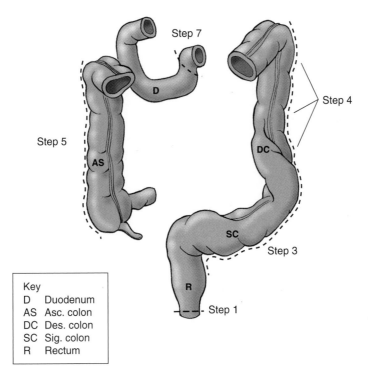

Step 7

Step 4

Step 5

Step 3

Step 1

D

AS

DC

SC

R

Key
D   Duodenum
AS Asc. colon
DC Des. colon
SC Sig. colon
R   Rectum

FIGURE 2.17 Incisions for GI removal.

5. Detach the ascending colon from the abdominal wall keeping *lateral* to its margin to preserve its vasculature.

6. Pull the transverse colon caudally, expose the origin of the SMA, and sever this vessel close to the aorta. Also cut the SMV and IMV at the inferior border of the pancreas. That portion of the pancreas posterior to the superior mesenteric vessels is known as the **uncinate process.**

7. Tie a double ligature around the duodenojejunal junction and cut the bowel between the two strings.

8. Sever the **suspensory muscle of the duodenum** (ligament of Treitz) where it tethers the duodenojejunal junction, and sever the radix (root) of the mesentery suspending the small bowel.

9. Carefully lift the detached GI tract out of the abdomen and place it on your table. Arrange the bowel in its characteristic "anatomical" configuration for further study.

C. **Detailed Examination of the Intestines**

1. **Jejunum and Ileum.** Open and clean a small portion of the jejunum and note the numerous **plicae circulares** (mucosal folds) lining the bowel's lumen. Open a small segment of the ileum about one foot from the terminal ileum. Note fewer plicae circulares and compare with the jejunum. Note that when compared to the ileum, the jejunum has a larger diameter, a thicker wall, greater vascularity, longer vasa recta, arterial arcades with larger loops, and fewer lymphoid nodules (Peyer's patches) in its mucosal wall. Now, open the cecum, clean it out, and look at the **ileocecal valve** and orifice of the appendix (these openings may be difficult to discern in the cadaver). If available, study x-rays of the barium-filled small intestine.   G2.63/C335/ R290/ N263, 265

2. **Colon.** Open a small portion of the transverse colon and note the **haustra** and **plicae semilunares.** If available, study the colon on barium-filled x-rays.   G2.63-2.64/C341-342/R287/ N267

Review the vascular supply to the bowel, again identifying the major branches of the SMA and IMA. Store the intestines in a plastic bag and keep for future reference.

D. **Unpaired Organs (Stomach, Spleen, and Liver)**

1. **Portal Vein.** First, trace the main tributaries of the portal vein, the **splenic vein, SMV,** and **IMV.** Look for **esophageal** and **gastric veins** and appreciate their important role in venous return to the heart in portal hypertension (Fig.2.12).   G2.49-2.50/C298/ R282/ N293

2. **Celiac Trunk.** Review the celiac trunk and its three branches, and note the autonomic fibers surrounding the celiac artery (celiac plexus).   G2.96/C344/ R314/N300-302

3. **Stomach.** Open the stomach along its greater curvature continuing your incision into the first part of the duodenum. Wash the mucosa with your sponge and identify **rugae, pyloric antrum** and **canal,** and the **pyloric sphincter** (Fig. 2.18). G2.24/C296/R276/ N258-260

4. **Spleen.** Note the **hilus,** where splenic vessels enter and leave the spleen. Study the relationships of the splenic vessels to the pancreas and note that the tail of the pancreas usually reaches the hilum of the spleen (Fig. 2.13).   G2.54/C300/ R282/ N281

E. **Removal of the Liver.** If your instructor would like the liver removed for further study, proceed as follows (Fig. 2.19):

1. Cut the falciform ligament between the diaphragm and diaphragmatic surface of the liver (dashed lines in Fig. 2.19). Cut the left triangular ligament and anterior portion of the coronary ligament. Then, cut the remaining portions of the coro-

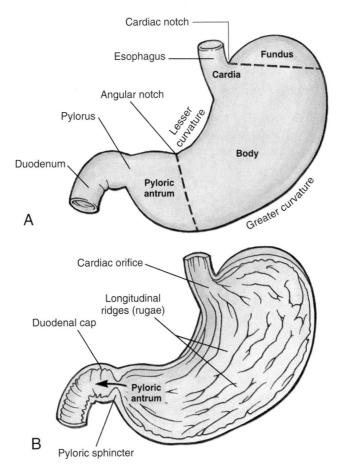

FIGURE 2.18 Stomach and proximal duodenum. A. External surface. B. Internal surface. Arrow passes through pyloric canal.

FIGURE 2.19 Anterior aspect of liver. Dash lines show cuts.

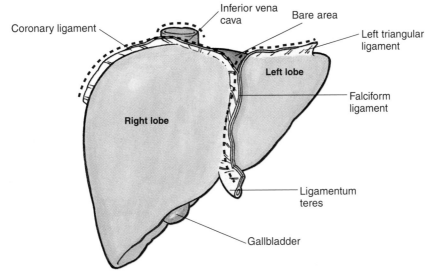

nary ligament. Now, pull the liver inferiorly and cut the inferior vena cava at the point where it pierces the diaphragm.

2. Pulling the liver anteriorly and superiorly, cut through the IVC as close to the liver as possible.

3. Sever the portal vein, common bile duct, and proper hepatic artery about midway along their course in the previously dissected hepatoduodenal ligament. Then carefully remove the liver, freeing any remaining attachments or adhesions.

F. Liver

1. Liver. Note the **bare area** of the liver and again identify the **coronary ligaments.** The bare area is where the liver is attached to the diaphragm and is not covered by visceral peritoneum. Observe the four "anatomic" lobes of the liver: **right, left, quadrate,** and **caudate** (Fig. 2.20). Functionally, the liver has only two lobes based on its internal blood supply. **G2.31/C309-311/ R280-281/ N270-272**

2. Study the **porta hepatis** and identify the **portal vein, hepatic ducts,** and **hepatic artery.** Find the fissure for the **ligamentum venosum,** the adult remnant of the embryonic ductus venosus. Examine the **IVC** and look for two or three large **hepatic veins** that drain directly into it (Fig. 2.20).

FIGURE 2.20   Posterior and inferior aspect of the liver.

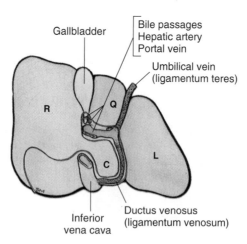

Key
R Rt. lobe
L Lt. lobe
C Caudate
Q Quadrate

VII. POSTERIOR ABDOMINAL STRUCTURES

Learning Objectives

- Describe the gross internal features of the kidneys and trace the pathway of urine from the renal pyramids to the urinary bladder.
- On a diagram, label the anterior visceral relationships of both kidneys.
- List the arterial supply to the adrenal glands.
- List and diagram the branches of the abdominal aorta and describe the structures or regions supplied by these branches.
- List and diagram the branches of the portal and caval venous systems and identify their important anastomoses.

Key Concepts

- Visceral relationships of the kidneys
- Vascular supply and venous drainage of abdomen
- Retroperitoneal

A. Introduction. The posterior abdominal structures are retroperitoneal and lie between the parietal peritoneum (covers the anterior surface of these structures) and the muscles and bones of the posterior abdominal wall.

B. Gonadal Vessels

1. Carefully remove the peritoneum from the posterior wall without damaging the small gonadal arteries. The **gonadal arteries** originate from the aorta inferior to the renal arteries. Observe that the **left gonadal (testicular or ovarian) vein** drains into the left renal vein, while the right gonadal vein drains into the IVC. Note that the gonadal vessels cross the external iliac vessels close to the ureters, and that the corresponding gonadal veins are larger than the gonadal arteries. **G2.77/C345/ R309-311/ N320**

C. Kidneys

1. Remove the kidneys from the renal fascia and their fatty capsule but leave the organs attached to their vessels and in the body cavity. Perirenal fat lies between a fibrous layer of renal fascia and the capsule of the kidney (Fig. 2.21). External to this renal fascia is a layer of pararenal fat found largely posterior to the kidney. This fatty "bed" permits the kidney to move during respiration. Clean the **left renal vein** from IVC and

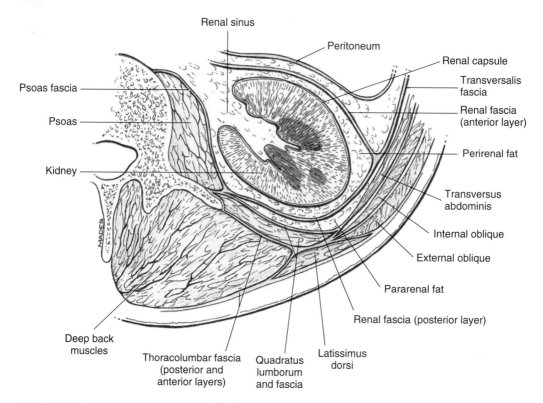

Renal sinus

Peritoneum

Renal capsule

Transversalis fascia

Renal fascia (anterior layer)

Perirenal fat

Psoas fascia

Psoas

Kidney

Transversus abdominis

Internal oblique

External oblique

Pararenal fat

Renal fascia (posterior layer)

Deep back muscles

Thoracolumbar fascia (posterior and anterior layers)

Quadratus lumborum and fascia

Latissimus dorsi

FIGURE 2.21   Transverse section of kidney showing its relationships to muscles and fasciae.

note that the **left gonadal vein** and venous tributaries from the suprarenal gland flow into the left renal vein. Now cut the left renal vein close to the IVC and identify the **renal artery.** Often, multiple renal arteries are present. Identify the **renal pelvis** and **ureter,** and carefully clean the ureter as it passes to the pelvis. Note that the ureter crosses the psoas muscle, passes obliquely posterior to the gonadal vessels, and that the proximal right ureter runs lateral to the IVC.   **G2.77, 2.95/C344-345/ R311-312/N311, 313-315**

2. Lift and reflect the duodenum to the left to reveal the right kidney. Clean the right renal vessels and right renal pelvis. The **right renal artery** passes posterior to the IVC. Reflect both kidneys to the midline by removing the substantial fatty renal capsule. Then, strip the fat and posterior renal fascia from the posterior abdominal wall and identify the **transversus abdominis, quadratus lumborum,** and **psoas major muscles.** Also note the **diaphragm** and **floating 12th rib.** In your atlas, study the anterior and posterior relationships of the kidneys. This is important in "reconstructing mentally" or "visualizing" the disposition of the abdominal viscera.   **G2.77/C345, 360/R310/ N247, 312**

3. The superior pole of the right kidney lies at the level of the 12th rib while the left kidney is somewhat higher. The liver

accounts for this difference, "pushing" the right kidney down and the dome of the diaphragm (and right lung) up.

D. Sectioned Kidney

1. Divide the left kidney into anterior and posterior halves by splitting the kidney along its lateral border. Do not sever the ureter or vessels.

Open the sectioned kidney like a book and identify the **renal cortex, medulla, renal pyramids, columns, papillae,** and the **minor and major calyces,** which unite to form the **renal pelvis.** The continuation of the pelvis is the ureter (Fig. 2.22). Observe that the ureter crosses the psoas muscle, passes posterior to the gonadal vessels, and then passes anterior to the bifurcation of the common iliac vessels to enter the pelvis. **G2.77, 2.80, 3.20/C355-356/ R306-307/ N313, 315**

E. Adrenal (Suprarenal) Glands

1. The adrenal glands lie just superior to the kidney. The **right adrenal gland** is usually triangular in shape and lies just posterior to the IVC. If the liver is not removed in your cadaver, then the right gland will be difficult to reach. The **left adrenal** is usually semilunar in shape. Postmortem, these glands deteriorate rapidly and may not be well preserved in your ca-

FIGURE 2.22   Kidney.

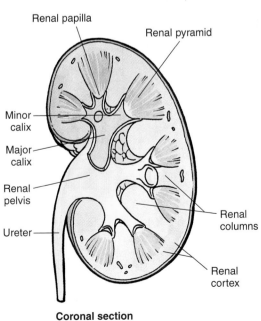

Coronal section

daver. Also, they may be embedded in fat and difficult to discern immediately. The arterial supply to the adrenal glands comes from the **inferior phrenic arteries, aorta,** and **renal arteries.** Section one gland and try to differentiate the cortex from the darker appearing medulla.   **G2.77, 2.95/C345-349/R309, 311/ N314, 325**

F. Abdominal Aorta and Inferior Vena Cava

1. Review the abdominal aorta and its branches (Fig. 2.23): three unpaired branches to the GI tract (celiac, SMA, IMA), three paired branches to paired organs (suprarenal, renal, gonadal), and branches to the walls of the abdomen (inferior phrenics, lumbars, median sacral). Identify one of the four pairs of **lumbar arteries.** Note that the aortic bifurcation occurs at the level of the L4 vertebral body. As a useful surface landmark, the umbilicus lies just superior to the aortic bifurcation.   **A3 1:55:29-2:02:27/ G2.100/C364/R309, 313/N320**

2. Also, review the IVC and its tributaries (Fig. 2.24). Note that the lumbar and ascending lumbar veins communicate with the azygos system of veins in the thorax, providing collateral venous return to the heart.

FIGURE 2.23   Arteries of posterior abdominal wall.

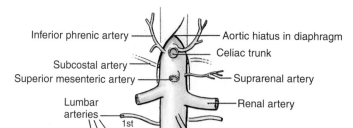

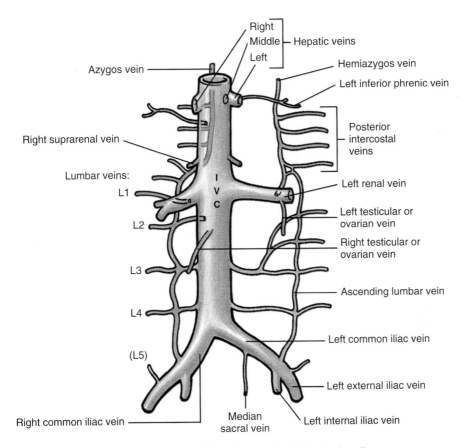

FIGURE 2.24 Veins of posterior abdominal wall.

VIII. POSTERIOR ABDOMINAL WALL

Learning Objectives

- Identify the muscles of the posterior abdominal wall and list their actions.
- List the nerves that form the lumbar plexus.
- Study the attachments and openings of the diaphragm.
- Review the autonomic innervation of the abdomen and define sites of referred pain from abdominal viscera.
- Identify key features of the abdominal viscera on radiographs and on cross sections of the abdomen.

Key Concepts

- Referred pain from abdominal viscera to T5-L2 dermatomes
- Lumbar plexus

A. Muscles of the Posterior Abdominal Wall. Muscles of the posterior wall include (Table 2.2) the following: **A3 1:31:11-1:47:15/ G2.91/C364-365/ R313-315/ N246**

1. **Psoas major.** Acting with the iliacus to flex the hip, it attaches to the lesser trochanter of the femur. ★ *flex hip*

2. **Iliacus.** Occupies the iliac fossa and with the psoas forms the iliopsoas, the most powerful flexor of the hip.

3. **Quadratus lumborum.** Extends and laterally flexes the vertebral column.

4. **Transversus abdominis.** Runs horizontally posterior to the quadratus lumborum.

B. Diaphragm

1. This skeletal muscle is innervated by the phrenic nerve (from C3, C4, and C5 spinal cord segments). Observe the sternal, costal, and lumbar attachments of the diaphragm (best observed in cadavers in which the liver has been removed). Identify the **right** and **left crus** (L, leg). Where the diaphragm arches over the aorta and posterior abdominal muscles, it forms "**arcuate ligaments.**" The medial arcuate ligament

TABLE 2.2
PRINCIPAL MUSCLES OF POSTERIOR ABDOMINAL WALL

Muscle	Superior Attachments	Inferior Attachment(s)	Innervation	Actions
Psoas major[a]	Transverse processes of lumbar vertebrae; sides of bodies of T12-L5 vertebrae and intervening intervertebral discs	By a strong tendon to lesser trochanter of femur	Lumbar plexus via ventral branches of L2–L4 nn.	Acting superiorly with iliacus, it flexes hip; acting inferiorly it flexes vertebral column laterally; it is used to balance the trunk when sitting; acting inferiorly with iliacus, it flexes trunk
Iliacus[a]	Superior two-thirds of iliac fossa, ala of sacrum, and anterior sacroiliac ligaments	Lesser trochanter of femur and shaft inferior to it, and to psoas major tendon	Femoral n. (L2–L4)	Flexes hip and stabilizes hip joint; acts with psoas major
Quadratus lumborum	Medial half of inferior border of 12th rib and tips of lumbar transverse processes	Iliolumbar ligament and internal lip of iliac crest	Ventral branches of T12 and L1–L4 nn.	Extends and laterally flexes vertebral column; fixes 12th rib during inspiration

[a]Psoas major and iliacus muscles are often described together as the iliopsoas muscle when flexion of the hip is discussed. The iliopsoas is the chief flexor of the hip, and when hip is fixed, it is a strong flexor of the trunk (e.g., during situps).

arches over the psoas major and the lateral arcuate ligament arches over the quadratus lumborum. The **median arcuate ligament** arches over the aorta in the midline. The central portion of the diaphragm is tendinous (**central tendon**) and the pericardial sac is attached to it superiorly. **G2.92/C362-363/R265, 315/N181, 246**

2. Identify the three large openings in the diaphragm: the **vena caval foramen** (at vertebral level 8), **esophageal hiatus** (vertebral level T10), and **aortic hiatus** (vertebral level 12) (also the opening for the thoracic duct and azygos vein) (Fig. 2.25). Look for the **greater splanchnic nerves** as they pierce the crura of the diaphragm and proceed to the **celiac ganglion** (Fig. 1.12, Chapter 2). To assist in finding this nerve, identify the greater splanchnic nerve in the thorax and push a probe through the diaphragm parallel to the nerve and then locate the probe on the abdominal side of the diaphragm. **G2.92,2.95/C364 R315/ N246, 250**

3. Review in your textbook and atlas the autonomic innervation of the abdominal viscera, how this innervation pattern relates to the blood supply of the embryonic GI tract, and understand the phenomenon of referred abdominal pain. **G2.95-2.97/C238-240/ R314/ N303, 306-307**

C. **Nerves of the Posterior Wall (Lumbar Plexus).** Remove the investing fascia of the posterior wall muscles to reveal the following nerves (on one side, feel free to remove a portion of the psoas major muscle to better expose these nerves) (Fig. 2.26): **A3 2:02:28-2:06:15/ G2.91/C364-365/R313, 447/ N250, 462-464**

1. **Subcostal (T12).** Just inferior to the 12th rib (not part of the lumbar plexus).

FIGURE 2.25 Diaphragmatic apertures.

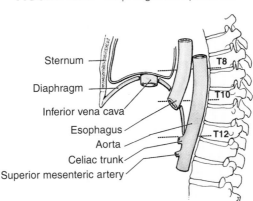

Sternum

Diaphragm

Inferior vena cava

Esophagus

Aorta

Celiac trunk

Superior mesenteric artery

T8

T10

T12

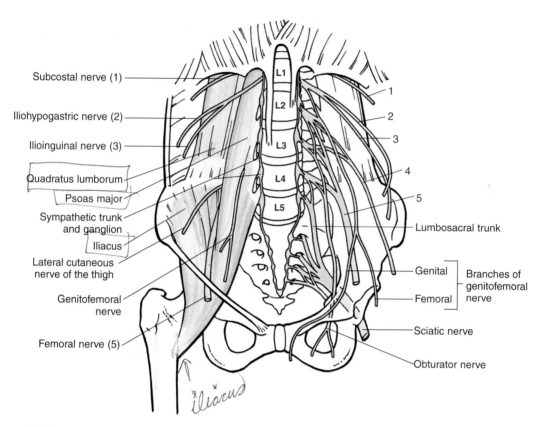

Subcostal nerve (1)

Iliohypogastric nerve (2)

Ilioinguinal nerve (3)

Quadratus lumborum

Psoas major

Sympathetic trunk and ganglion

Iliacus

Lateral cutaneous nerve of the thigh

Genitofemoral nerve

Femoral nerve (5)

L1
L2
L3
L4
L5

1
2
3
4
5

Lumbosacral trunk

Genital

Femoral

Branches of genitofemoral nerve

Sciatic nerve

Obturator nerve

iliacus

FIGURE 2.26 Posterior abdominal wall. Muscles removed on the right to demonstrate the nerves. Numbers refer to nerves on the right side.

2. Iliohypogastric and ilioinguinal nerves (L1). Descend anterior to the quadratus lumborum and frequently arise from a common trunk that may not separate until it reaches the transversus abdominis muscle.

3. Genitofemoral (L1-2). Pierces the anterior surface of the psoas muscle. It divides into two branches (genital and femoral) and supplies skin of the medial thigh just inferior to the inguinal ligament and innervates the cremaster muscle.

4. Lateral cutaneous nerve of the thigh (L2-3). Passes deep to the inguinal ligament near the anterior superior iliac spine and is sensory to the lateral thigh.

5. Femoral (L2-4). Large nerve that lies between the psoas and iliacus and then passes deep to the inguinal ligament. Its motor fibers innervate the extensors of the knee (anterior thigh muscles).

6. Obturator (L2-4). Medial to the psoas and deep; passes through the obturator foramen to provide motor fibers to the adductor muscles of the hip (medial thigh muscles).

7. **Lumbosacral trunk (L4, 5).** Descends into the pelvis to become part of the sacral plexus. Difficult to see unless the psoas is partially removed. It will form part of the large sciatic nerve in the deep pelvis.

8. Although not part of the lumbar plexus, identify the **sympathetic trunk** closely applied to the lateral sides of the lumbar vertebrae and look for rami communicantes passing from sympathetic ganglia to lumbar somatic nerves.

9. Review the disposition of the abdominal viscera in cross sections of the abdomen.   **G2.103/C370-375/ R300-301, 304/ N517-522**

— lumbosacral trunk
is part of
large sciatic nerve

3

PELVIS AND PERINEUM

I. PELVIS

Learning Objectives

- Throughout Chapter 3, identify structures in bold print unless instructed to do otherwise.
- Identify features of the bony pelvis and define the terms "greater" and "lesser" pelvis.
- Compare and contrast the female and male pelvis.
- Identify the muscles that contribute to the "pelvic diaphragm" and their actions.
- Describe the peritoneal relationships of the pelvis in both sexes.

Key Concepts

- Sex differences in the bony pelvis
- Difference between pelvic and abdominal diaphragms
- Subdivisions of the perineum

A. General Remarks and Definitions

1. This laboratory approach includes an overview of the pelvic peritoneal reflections, identification of the major pelvic viscera, a brief dissection of the anal triangle, a more detailed dissection of the urogenital triangle, and hemisection of the pelvis into right and left halves. Depending on the approach at your institution, you may want to modify this sequence. Because of the difficulty in dissecting the perineum and appreciating its three-dimensional anatomy, you

should make use of pelvic models and other study aids as you review this clinically important region. First, begin by observing the following features. No dissection is required at this time.

2. The pelvis (L. pelvis, basin) is divided into the greater and lesser pelvis. The **greater (major) pelvis** is located superior to the **pelvic brim** and is bounded on either side by the ilium (Fig. 3.1). The **lesser (minor or true) pelvis** is situated inferior to the pelvic brim.   G3.1/ R408-414/ N330-332

3. The walls of the pelvic cavity are partially lined with muscles. The floor of the pelvis is formed by muscles collectively called the **pelvic diaphragm** (levator ani and coccygeus muscles). G3.22/C418/ R324-325, 328/ N333-336

4. The rectum passes through the pelvic floor to the **anal triangle.** The urinary and genital system pass through the pelvic floor to the **urogenital triangle (UG triangle)**. The **perineum** is a diamond-shaped area extending from the symphysis pubis to the coccyx. A transverse line between the right and left ischial tuberosities divides the perineal region into two triangular areas, the UG triangle and the anal triangle, mentioned above.   G3.37/C421-426/R328-331/ N354

5. The dissections of the male and female pelvis will be covered separately. However, you need to learn the anatomy of this important region in both sexes, so be sure to study the pelvis of the opposite sex in a cadaver at another table.

FIGURE 3.1   Lower abdominopelvic cavity. *Green,* pelvis major; *red,* pelvis minor; *broken line,* plane of pelvic brim surrounding the superior pelvic aperture.

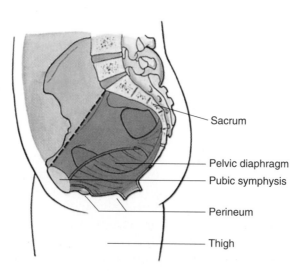

Sacrum

Pelvic diaphragm

Pubic symphysis

Perineum

Thigh

B. Important Landmarks (Both Sexes)

1. The bony pelvis is formed by the **os coxae** (hip bones), the sacrum, and coccyx.   **A3 2:07:57-2:15:57/ G3.1, 3.3/C386-389/R411-414/N330-331**

2. **Os coxae** (hip bone) consists of three parts: **ilium, ischium,** and **pubis.** All three fuse at the **acetabulum.**

3. **Sacrum** (L. sacer, sacred). Identify the **promontory, sacral canal,** and **anterior** and **dorsal sacral foramina** (for passage of ventral and dorsal rami of spinal nerves).

4. **Coccyx** (Gr. kokkyx, cuckoo; resembles a cuckoo's bill). Three to five rudimentary fused vertebrae.

5. The **pelvic brim** surrounds the pelvic inlet (superior aperture).

6. Identify the **obturator foramen, ischial tuberosity, ischial spine, pubic arch,** and, on models if available, the **sacrospinous ligament** and **sacrotuberous ligament.**

7. The sacrospinous and sacrotuberous ligaments partially bound two foramina: **lesser sciatic foramen** and **greater sciatic foramen.**

C. Observations of the Female Pelvis Minor (True Pelvis)

1. On the inside aspect of the anterior abdominal wall again find and review the location of the **median umbilical ligament** (remnant of the embryonic urachus) and **medial umbilical ligament** (obliterated umbilical artery of fetus) (Fig. 3.2). Identify where the **ureter** and **ovarian blood vessels** cross the pelvic brim. Find the **round ligament of the uterus** and note that it crosses the external iliac vessels near the inguinal ligament and enters the deep inguinal ring (Fig. 3.3).   **G3.27, 3.34/C262/ R332-338/ N236, 339**

2. The peritoneum reflects over the superior surface of the bladder and then onto the uterus. Between the bladder and uterus lies the **vesicouterine pouch.** Between the uterus and the rectum, the peritoneum forms the deep **rectouterine pouch** (of Douglas). Then, peritoneum reflects onto the anterior surface of the rectum. Pararectal and paravesical fossae lie on either side of the rectum and bladder where the peritoneum reflects onto the walls of the pelvis.   **G3.26,3.28/C394,398/R335/N339-342**

3. The uterus is almost entirely covered by peritoneum. The **broad ligament** is a fold of peritoneum that stretches on each side of the uterus to the side walls of the pelvis. In its free su-

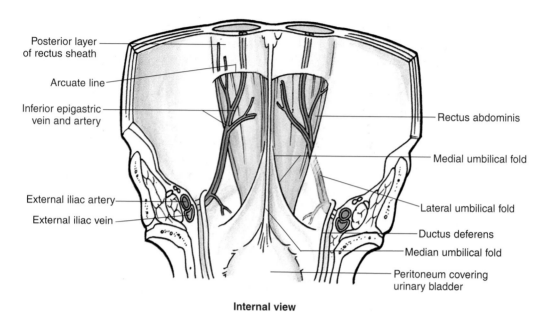

Internal view

FIGURE 3.2 Internal view of anterior abdominal wall.

FIGURE 3.3 Female pelvic viscera viewed from above looking into the pelvis.

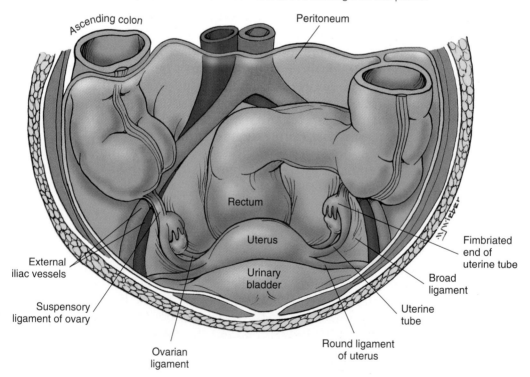

perior border lies the **uterine tube** (Fallopian tube). Note that the broad ligament is really two layers of peritoneum, similar to a mesentery. Identify the **mesovarium,** which supports the ovary, the **mesosalpinx,** which surrounds the uterine tube, and realize that both of these subdivisions are really all part of the broad ligament. The lateral extension of the broad ligament from the ovary to the lateral pelvic wall is the **suspensory ligament of the ovary** (and contains the ovarian vessels). At the base of the uterus, the **rectouterine folds** appear as sharp peritoneal folds that contain ligaments, the **sacrouterine ligaments,** which help support the uterus and anchor it to the sacrum. Note from your atlas and textbook that the pelvic fascia in females is thickened adjacent to the cervix and vagina. Since these thickened fascial ligaments lie beneath the peritoneum, they are difficult to visualize but form the important **transverse cervical ligaments** (cardinal ligaments, or ligament of Mackenrodt) which help support the uterus and anchors it to the lateral pelvic walls (Fig. 3.4). **G3.27, 3.32/C399-405/ R333-337/ N344-346**

4. The **uterine tube** usually turns downward and posterior so that its **fimbriated end** comes into close relationship with the ovary.

5. Posteriorly, note that the fused sacral vertebrae S3 to S5 and the coccyx are covered anteriorly by the rectum. Passing two fingers behind the rectum at this level, your fingers come to lie in the retrorectal space. This space is limited inferiorly by the strong fascia covering the levator ani muscle (part of the pelvic diaphragm). As you move your fingers laterally to the sides of the retrorectal space, you feel

FIGURE 3.4 Ligaments of pelvic floor (female).

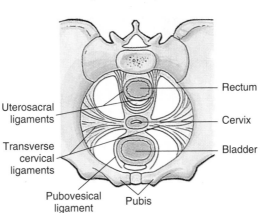

strands of the pelvic splanchnic nerves (parasympathetics from S2-4) bilaterally. These fibers eventually converge on the **pelvic plexus (inferior hypogastric plexus)** covering both sides of the rectum, uterus, vagina, and bladder just deep to the peritoneum.

Study the male cadaver before leaving lab.

D. Observations of the Male Pelvis Minor (True Pelvis)

1. On the inside aspect of the anterior abdominal wall again find and review the location of the **median umbilical ligament** remnant of the embryonic urachus).

Laterally, identify the **medial umbilical ligament,** a cordlike structure representing the obliterated umbilical artery of the fetus (Fig. 3.2). The ureter crosses the pelvic brim at about the point where the common iliac arteries divide into internal and external branches (an important relationship). Find the **ductus deferens** as it crosses the external iliac vessels. **G3.14/C262, 286/ R316-319/ N236, 340**

2. The peritoneum within the male pelvis covers the superior surface of the bladder and then reflects upward onto the lateral pelvic wall. Posterior to the bladder, the peritoneum reflects downward and then upward onto the anterior surface of the rectum. Between the bladder and rectum lies the **rectovesical pouch.** Paravesical and pararectal fossae lie on each side of the bladder (better seen when the bladder is full) and rectum, respectively. Realize, from sagittal views of the pelvis, that the seminal vesicles and prostate gland lie deep to the peritoneum of the rectovesical pouch and are not visible at this time (Fig. 3.12 later in this chapter). **G3.7/C439-440/ R316/N338, 340**

3. Posteriorly, note that the fused sacral vertebrae S3 to S5 and the coccyx are covered anteriorly by the rectum. Passing two fingers behind the rectum at this level, your fingers come to lie in the retrorectal space. This space is limited inferiorly by the strong fascia covering the levator ani muscle (part of the pelvic diaphragm). As you move your fingers laterally to the sides of the retrorectal space, you feel strands of the **pelvic splanchnic nerves (parasympathetics from S2-4)** bilaterally. These fibers eventually converge on the **pelvic plexus (inferior hypogastric plexus)** covering both sides of the rectum and bladder just deep to the peritoneum.

Study the female cadaver before leaving lab.

II. ANAL AND UROGENITAL TRIANGLES

Learning Objectives

- Identify the boundaries of the perineum and its subdivisions.
- Describe the features of the external genitalia in both sexes.
- On a diagram, label the muscles, vessels, and nerves found in the UG triangle.
- List the muscles in the deep perineal space.
- Diagram the extent of the pudendal nerve and its major perineal branches.

Key Concepts

- Ischioanal fossa and anal sphincters
- Perineal pouches and spaces
- Deep perineal space (UG diaphragm)

A. Anal Triangle (Both Sexes)

1. With the cadaver in the supine position, spread the legs as far apart as possible using a wooden block or place the legs in stirrups, if available. Incise the skin and reflect it, as shown in Figure 3.5 (females) or Figure 3.6 (males) to expose the perineal region. If the anal region is too difficult to reach from this aspect, turn the cadaver to the prone position. Before proceeding further, stabilize and distend the anal canal by inserting a super-size tampon, along with its plastic insertion tube, into the anal canal. This adds rigidity to the anus and will make the dissection easier.

2. After removing the skin and tela (superficial fascia), clean away the large amount of fat which occupies the **ischioanal (ischiorectal) fossa** (accommodates distension of the lower rectum) and using blunt dissection identify **the inferior rectal nerves** and **vessels** passing to the anal canal. Note that the nerves come from the pudendal nerve (S2-4) and the vessels are branches of the **internal pudendal artery** and **vein** which traverse the pudendal canal as they course anteriorly to the perineum.   **G3.39/C465/ R342-343/ N382**

3. Clean the **external anal sphincter,** noting its anterior attachment to the **perineal body** (central tendon of the perineum) (Figs. 3.5 and 3.6). Once the ischioanal fossa is cleared of fat, note the deep aspect of the levator ani muscle (pelvic diaphragm) lateral to the anal sphincter.

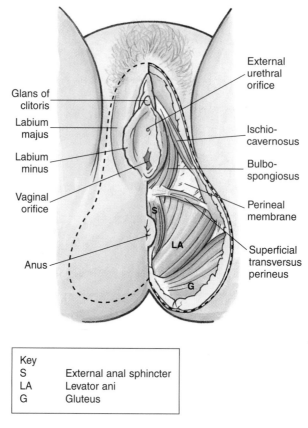

Key	
S	External anal sphincter
LA	Levator ani
G	Gluteus

FIGURE 3.5   Female perineum. Dash lines show skin incisions

B. General Features of the UG Triangle

1. The urogenital (UG) triangle is the anterior division of the perineum. Passing through this region are the terminal portions of the urinary and genital systems. Superficial fascia (Colles' fascia) covers the triangle and is a continuation of the membranous layer of superficial fascia (Scarpa's fascia) seen in the abdominal wall, and is continuous with the dartos fascia of the penis and scrotum. Inferiorly, this fascia has strong attachments to the inguinal ligament, ischiopubic rami, and posteriorly blends with the perineal body. Between this superficial fascia and the perineal membrane (inferior fascia covering the sphincter urethrae muscle) lies the **superficial space.** The **deep space** is composed of the sphincter urethrae and deep transverse perineal muscles, their fascial coverings, and small glands.   **G3.37, 3.39/C421-431/ R325/ N351-352**

C. Female External Genitalia

1. Place the cadaver in the supine position with the legs spread far apart (see above). Complete the skin incisions

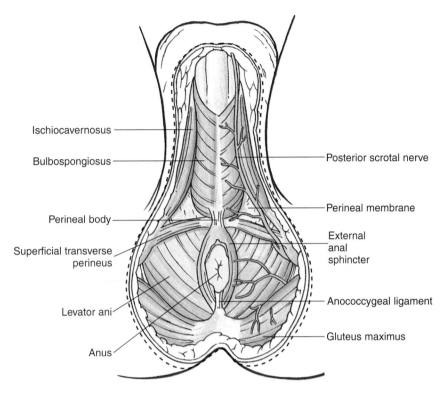

FIGURE 3.6   Male perineum. Dash lines show skin incisions.

shown in Figure 3.5, including removal of the skin from the labia majora.

2. In the female, the genital swellings of the fetus did not fuse so a midline separation called the **vestibule** exists which transmits the urethra, vagina, and ducts of the greater vestibular glands. The vestibule is bounded by the **labia majora** and **minora.** Anteriorly, the labia majora blend together to form the **mons pubis.** The labia minora divide anteriorly to form the **prepuce of the clitoris** and posteriorly to form the **frenulum.** G3.47-3.50/C426/ R340-341/ N350

3. Remove the skin of the **clitoris,** identify the **corpora cavernosa,** and follow them into the superficial space along the ischiopubic rami. The crura are covered by thin **ischiocavernosus muscles.** The small **glans clitoris** capping the corpora cavernosa is formed by the anterior ends of the cavernous bodies, which join to form the **body of the clitoris.** G3.48/C430/ R341-344/ N351-353

4. Flush with the vaginal opening, remove the two labia minora, and clean the dissection field thoroughly to expose the **bulb of the vestibule.**

5. Because midline fusion did not occur in the female, the **bulb of the vestibule** is divided into two erectile tissue bodies covered by the **bulbospongiosus muscle.** Attached to the posterior end of the bulb are the **greater vestibular glands** (Bartholin's glands). The **external urethral orifice** is situated in the vestibule between the clitoris and vagina. Cut the bulbospongiosus muscle and remove it to see the bulbs of the vestibule (two erectile tissue masses on each side of the vaginal opening). Attached to the posterior end of each bulb is the greater vestibular gland (Bartholin's glands) which may be difficult to identify in the cadaver.

6. The **perineal body** in the female is located between the anus and vagina and is the attachment point for the external anal sphincter, bulbospongiosus muscle, and **superficial transverse perineal muscle** (often absent). It forms an important support for pelvic viscera in the female. The perineal membrane spans the ischiopubic rami, forming the inferior fascia of the deep perineal pouch. If possible, incise this fascial plane and view the structures of the deep perineal space. **G3.50/C428-430/ R341/ N352, 367**

7. The **deep perineal space** is difficult to dissect, and a thorough understanding of this area should be obtained from your textbook, atlas, and models. The deep space is occupied by several muscles and their fascial coverings (older terminology refers to these muscles collectively as the urogenital diaphragm). The **external sphincter urethrae muscle** surrounds the urethra as it passes through this space. Other muscles of the deep perineal space include the **deep transverse perineal muscle** (this muscle is very small or absent), the **compressor urethrae muscle** (part of the sphincter urethrae muscle), and the **urethrovaginal sphincter muscle** (most textbooks fail to identify these muscles as distinct structures). Because the vagina and urethra pass through the deep perineal space, these muscles are difficult to discern. These muscles are innervated by branches of the perineal nerve, which is a branch of the **pudendal nerve.** Blood vessels in this area are branches of the **internal pudendal vessels.** Study their distribution in your atlas. **G3.46-3.49/C427/ R342-343/N352-353, 375, 384**

Study the male urogenital triangle.

D. Male External Genitalia

1. Place the cadaver in the supine position with the legs far apart. Complete the skin incisions shown in Figure 3.6. The fatty layer of superficial fascia is continuous with the dartos muscle of the scrotum and the subcutaneous fatty layer on the medial aspect of the thighs. The membranous layer (Colles' fascia) is more aponeurotic and is continuous with the dartos fascia and the membranous layer of the abdominal superficial fascia (Scarpa's fascia). Colles' fascia is firmly attached to the pubis and ischium.   **G3.6/C445/ R316-317/ N354**

2. Make a midline incision along the ventral surface of the **penis** (L. penis, tail) and reflect the skin from the shaft. Identify the **glans penis, superficial dorsal vein** (drains into external pudendal veins of lower abdominal wall and thigh), and **deep fascia of the penis.** The deep fascia encircles the cavernous bodies. On the dorsal midline identify the **deep dorsal vein of the penis** (drains into the prostatic venous plexus within the pelvic cavity). Incise the deep fascia encircling the cavernous bodies and identify the **two corpora cavernosa** and **ventral corpus spongiosum,** which transmits the penile (spongy) urethra. Separate these cavernous bodies. Identify the paired dorsal arteries and nerves of the penis.   **G3.41-3.42/C460-462/R316-320, 326-328/ N355-356**

3. Identify the roots of the corpora cavernosa which form the **two crura** of the penis (each crus attaches to the **ischiopubic ramus**). Clean the **ischiocavernosus muscles** that surround these crura (Fig. 3.7). The root of the corpus spongiosum forms the bulb of the penis and is covered by the **bulbospongiosus muscle.** This muscle arises from the **perineal body** (central tendon of perineum).   **G3.39/C464/ R328-331/ N355**

4. From the perineal body centrally, look for the two halves of the **superficial transverse perineal muscle** extending to the

FIGURE 3.7 Deep dissection of the male perineum.

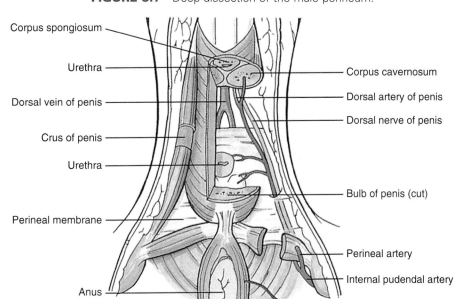

Corpus spongiosum

Urethra

Dorsal vein of penis

Crus of penis

Urethra

Perineal membrane

Anus

Corpus cavernosum

Dorsal artery of penis

Dorsal nerve of penis

Bulb of penis (cut)

Perineal artery

Internal pudendal artery

right and left ischial tuberosities. The perineal body is a dense connective tissue region that is the attachment point for muscle fibers of the superficial transverse perineus, bulbospongiosus, and external anal sphincter muscles. **G3.39/C464-465/ R331/ N367**

5. The **deep perineal space** is best studied from the atlas, textbook, and models as this area is difficult to dissect in the cadaver. The deep space in males is composed of several muscles (often referred to as the urogenital diaphragm), their fascial coverings, and the bulbourethral glands. Incise the perineal membrane, clean and identify the **external sphincter urethrae muscle** (a small **compressor urethrae muscle** forms part of this structure) surrounding the **membranous urethra** as it courses toward the bulb of the penis. Also, try to identify the **deep transverse perineal muscle** (this muscle is very small or absent). Two **bulbourethral** (Cowper's) **glands** are embedded in the sphincter urethrae muscle on either side of membranous urethra. Review the distribution of the perineal nerves and vessels in your atlas. **G3.37-3.38/C462/ R331/ N357-358, 376, 382**

6. Because many of the nerves and vessels in the perineum are small, it is important to read about their distribution in your textbook. Realize that the blood supply to the perineum is largely via the **internal pudendal artery** and its branches, while the somatic innervation is via the **pudendal nerve.** Pudendal means "to be ashamed," and the perineum evoked this name in early medical history.

Be sure to review the female urogenital triangle.

III. SPLIT PELVIS

Learning Objectives

- Identify the major branches of the internal iliac artery.
- Identify the female and male pelvic viscera and be able to describe their innervation.
- Be able to trace sperm from the testis to the spongy urethra naming in correct order each duct or gland traversed.
- Functionally describe micturition and ejaculation.
- List the ligaments and muscles that offer support for the uterus.

Key Concepts

- Importance of anal sphincter and urethral sphincter, and perineal body
- Internal iliac artery distribution
- Micturition

A. Introduction. Hemisection of the pelvis offers a unique opportunity to view many structures that lie deep within the pelvis and affords a chance to trace nerves and vessels. Check with your instructor to determine if you should complete this dissection.

B. Split Pelvis in the Female

1. The pelvis is to be split so that the anatomy of the pelvic viscera may be studied in sagittal section. Place the cadaver in the supine position. Cut through the midline of the pubic symphysis with a saw and then extend the cut posteriorly through the pelvic viscera (bladder, vagina, uterus, and rectum) to the sacrum *using a sharp scalpel.* Keep as close to the midline as possible.

Inferior to the pubic symphysis, make a midsagittal cut with a scalpel directed posteriorly and cut through the pelvic diaphragm to the coccyx.

Cut the right common iliac vessels at the level of the fourth lumbar vertebra. Also cut the right ureter near the renal pelvis and the right ovarian vessels. Reflect the aorta and inferior vena cava to one side and make a saw cut through the coccyx, sacrum, and lumbar vertebrae. Make a clean transverse cut through the soft tissues above the right iliac crest to the intervertebral disc between the third and fourth lumbar vertebrae. Now remove the right lower quadrant and lower extremity for further study (Fig. 3.8).

2. Rectum. The lowest portion of the rectum runs downward and backward almost at right angles to the upper part. It passes through the pelvic diaphragm to reach the **anal canal** (Fig. 3.9). Try to identify the anal columns and semilunar folds called the **anal valves** (these features often are difficult to discern on the cadaver, especially if the rectum is distended). Trace the **superior rectal artery** along the posterior surface of the rectum where it anastomoses with the middle rectal arteries from the internal iliac artery. Review the anal sphincter musculature.   **G3.11/C450/R335/ N365-366, 369-370**

3. Urethra and Urinary Bladder. If the hemisection was done perfectly in the midline, then the longitudinally opened halves of the urethra should be clearly visible. Verify that the **female urethra** is short (3.4 to 4 cm), and examine that portion of the urethra that passes through the deep perineal space and is surrounded by the **external sphincter urethrae muscle.** Inside the **urinary bladder** the mucous membrane lining forms irregular folds except in the area of the **trigone** which is smooth. The trigone demarcates the two ureteral orifices and the urethral orifice on the bladder's posterior wall (sometimes

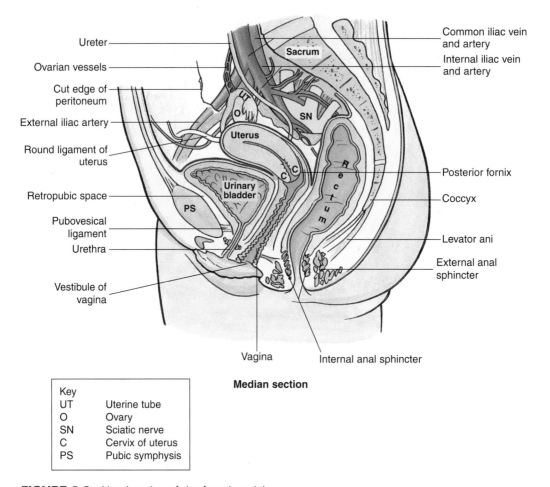

Ureter

Ovarian vessels

Cut edge of peritoneum

External iliac artery

Round ligament of uterus

Retropubic space

Pubovesical ligament

Urethra

Vestibule of vagina

Sacrum

Common iliac vein and artery

Internal iliac vein and artery

UT

O

SN

Uterus

C C

R e c t u m

PS

Urinary bladder

Posterior fornix

Coccyx

Levator ani

External anal sphincter

Vagina

Internal anal sphincter

Median section

Key	
UT	Uterine tube
O	Ovary
SN	Sciatic nerve
C	Cervix of uterus
PS	Pubic symphysis

FIGURE 3.8   Hemisection of the female pelvis.

FIGURE 3.9   Median section of the rectum and anal canal.

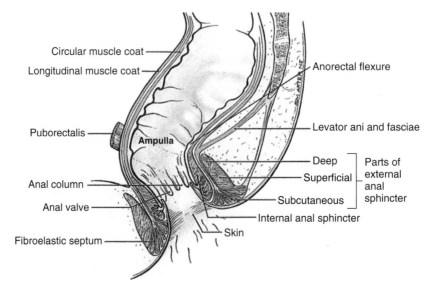

Circular muscle coat

Longitudinal muscle coat

Puborectalis

Anal column

Anal valve

Fibroelastic septum

Ampulla

Anorectal flexure

Levator ani and fasciae

Deep
Superficial
Subcutaneous

Parts of external anal sphincter

Internal anal sphincter

Skin

these openings are difficult to discern). The bladder wall is composed of the detrusor muscle (L. detrudere, to thrust out) (smooth muscle).   **G3.16-3.17/C442-443/R318, 332/ N343, 353**

4. **Vagina and Uterus.** Internally, transverse ridges of the vaginal wall form the **rugae vaginales.** Note that the uterus projects downward into the upper portion of the vagina where it opens by a small circular or oval aperture, the **external os.** Portions of the vagina extending above this lower portion of the uterus are known as the **posterior** and **anterior fornices of the vagina.** Note the relationship of the **posterior fornix** to the rectouterine pouch.   **G3.26,3.32/C394,400/R337-338/N345-346**

   In the uterus, note that there is a narrowing of the lumen just above the internal ostium (os) which separates the **body** of the uterus from the **cervix.** The mucosa lining the uterus is the **endometrium.** The smooth muscle wall of the uterus is called the **myometrium,** and the rounded portion of the uterus above the entrance of the uterine tubes is the **fundus.**

   Identify the **ovarian ligament,** a fibrous band that extends from the medial side of the ovary to the lateral border of the uterus (Fig. 3.3).

   Finally, again note the thickened pelvic fascia at the level of the cervix, which forms the **transverse cervical ligaments** (cardinal ligaments) (Fig. 3.4). While difficult to visualize, you may be able to feel these ligaments with your fingers. These important support ligaments anchor the uterus to the lateral pelvic walls.   **G3.34/C419/ N341**

5. **Arteries.** The **internal iliac (hypogastric) artery** gives rise to visceral branches, which include the **umbilical, vesical, uterine,** and **middle rectals** (Fig. 3.10). Clean and follow these branches to the anterior abdominal wall, bladder, uterus, and rectum, respectively. Feel free to remove veins that may obscure your field of dissection. Parietal branches (supply skeletal muscles) include the iliolumbar, lateral sacral, obturator, internal pudendal, and superior and inferior gluteal arteries. The **superior gluteal, inferior gluteal,** and **internal pudendal** usually arise in that order from the main portion of the internal iliac artery. Clean these three branches and note that they leave the pelvis via the greater sciatic foramen. Identify the **obturator artery** and follow it as it leaves the pelvis with the obturator nerve to enter the medial thigh (Fig. 3.10).   **A3 2:27:16-2:30:39/ G3.20, 3.32/C410-412/ R324-325/N373**

6. **Pelvic Diaphragm.** Push the pelvic viscera medially and clean the muscles of the pelvic wall and floor. Find the branches of the 2nd, 3rd, and 4th sacral nerves and the **leva-**

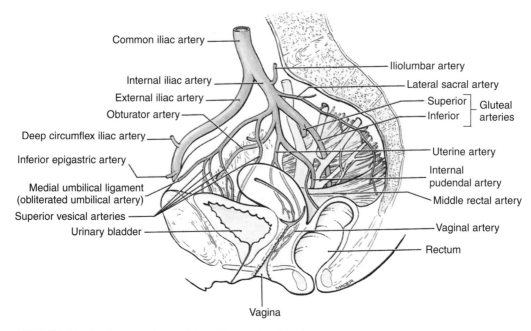

Common iliac artery

Internal iliac artery

External iliac artery

Obturator artery

Deep circumflex iliac artery

Inferior epigastric artery

Medial umbilical ligament
(obliterated umbilical artery)

Superior vesical arteries

Urinary bladder

Iliolumbar artery

Lateral sacral artery

Superior ⎤ Gluteal
Inferior ⎦ arteries

Uterine artery

Internal
pudendal artery

Middle rectal artery

Vaginal artery

Rectum

Vagina

FIGURE 3.10 Iliac arteries and branches in the female.

tor ani muscle (Fig. 3.11). Posteriorly, find the **coccygeus muscle** (often blends with the sacrospinous ligament) inserting into the coccyx and lower sacrum. Looking beneath sacral nerve segments S2-4, find the **piriformis muscle** arising from the sacrum. Its muscle fibers converge and pass through the greater sciatic foramen. **A3 2:15:58-2:27:15/ G3.21-3.22/C418/R342/ N333-334, 465**

Now review the male pelvis.

FIGURE 3.11 Lateral wall of the true pelvis showing pelvic diaphragm and its relationship to sacral and coccygeal plexuses.

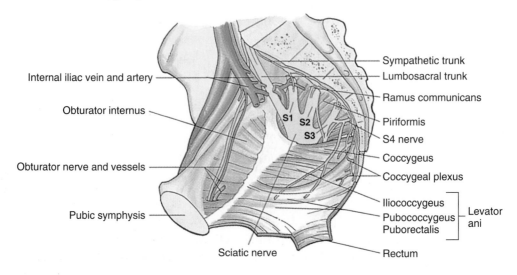

Internal iliac vein and artery

Obturator internus

Obturator nerve and vessels

Pubic symphysis

Sciatic nerve

Sympathetic trunk

Lumbosacral trunk

Ramus communicans

Piriformis

S4 nerve

Coccygeus

Coccygeal plexus

Iliococcygeus ⎤
Pubococcygeus ⎟ Levator
Puborectalis ⎦ ani

Rectum

S1 S2
S3

C. Split Pelvis in the Male

1. The pelvis now is to be split into two parts to study the viscera in a sagittal view. Place the cadaver in the supine position. Insert a probe into the spongy urethra and with a *sharp scalpel* bisect the corpus spongiosum by cutting down to the probe from both the dorsal and ventral sides.

Cut through the midline of the pubic symphysis with a saw and extend the cut posteriorly through the pelvic viscera (bladder, prostate, and rectum) to the sacrum using a sharp scalpel. Stay as close to the midline as possible.

Inferior to the pubic symphysis, make a midsagittal cut with a scalpel directed posteriorly and cut through the pelvic diaphragm to the coccyx.

Sever the right common iliac vessels at the level of the fourth lumbar vertebra. Then cut the right ureter near the renal pelvis and tie it to the right testicular vessels, which should be cut near the kidney. Reflect the aorta and inferior vena cava to one side and make a saw cut through the coccyx, sacrum, and lumbar vertebrae. Make a clean transverse cut through the soft tissues above the iliac crest on the right side to the intervertebral disc between the third and fourth lumbar vertebrae. Remove the lower right quadrant (pelvis and lower extremity) for further study (Fig. 3.12).

2. Rectum. The lowest portion of the rectum runs downward and backward almost at right angles to the upper part. It passes through the pelvic diaphragm to reach the **anal canal** (Fig. 3.9). Try to identify the **anal columns** and semilunar folds called the **anal valves** (these features often are difficult to discern on the cadaver, especially if the rectum is distended). Trace the **superior rectal artery** along the posterior surface of the rectum where it anastomoses with the middle rectal arteries from the internal iliac artery. Review the anal sphincter musculature.   G3.11/C450/R335/N365-366, 369-370

3. Bladder and Urethra. The **bladder** lies above the **prostate** and is in direct contact with the anterior surfaces of the seminal vesicles (Figs. 3.12 and 3.13). The **ureter** passes *under* the **ductus deferens** on its way to the bladder. The mucous membrane lining of the bladder forms irregular folds except in the area of the **trigone,** which is smooth. The trigone demarcates the two ureteral orifices and the urethral orifice on the posterior bladder wall (sometimes these openings are difficult to discern). The bladder wall is composed of the detrusor muscle (smooth muscle).   G3.16-3.17/C442-443/ R317-318/ N358-359

The **male urethra** is divided into three parts: **prostatic, membranous,** and **spongy (penile).** Observe that the membranous urethra passes through the deep perineal space and is only

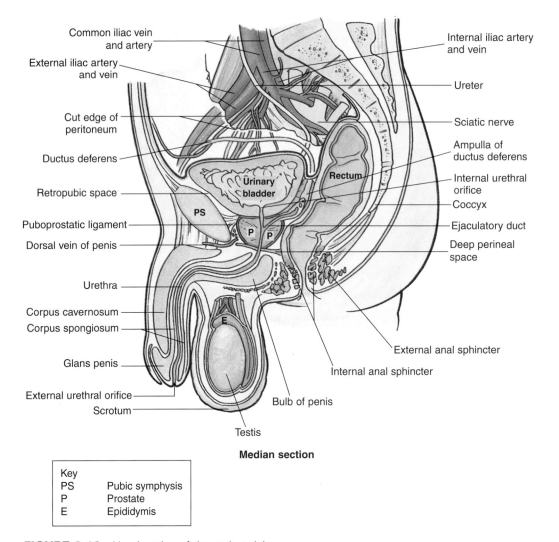

Common iliac vein
and artery

Internal iliac artery
and vein

External iliac artery
and vein

Ureter

Cut edge of
peritoneum

Sciatic nerve

Ductus deferens

Ampulla of
ductus deferens

Internal urethral
orifice

Retropubic space

Coccyx

Puboprostatic ligament

Ejaculatory duct

Dorsal vein of penis

Deep perineal
space

Urethra

Corpus cavernosum

Corpus spongiosum

External anal sphincter

Glans penis

Internal anal sphincter

External urethral orifice

Bulb of penis

Scrotum

Testis

Urinary bladder

Rectum

PS

P   P

E

Median section

Key
PS   Pubic symphysis
P   Prostate
E   Epididymis

FIGURE 3.12   Hemisection of the male pelvis.

about 1 cm long. Here it is surrounded by the **external sphincter urethrae muscle** (Fig. 3.13). From your textbook, be sure you understand how the muscles of the bladder and urethra function in micturition.

4. **Prostate and Seminal Vesicles.** The **prostate gland** is enclosed in a strong sheath and lies just above the deep perineal space and external sphincter urethrae muscle (Figs. 3.12 and 3.13). The deep dorsal vein of the penis enters beneath the pubic symphysis and above the deep perineal space to join the prostatic plexus of veins (difficult to sometimes appreciate in the cadaver).

The **seminal vesicles** are single long tubes folded upon themselves. The ducts of the seminal vesicles are joined by the ductus (vas) deferens to form the **ejaculatory ducts.** The vas def-

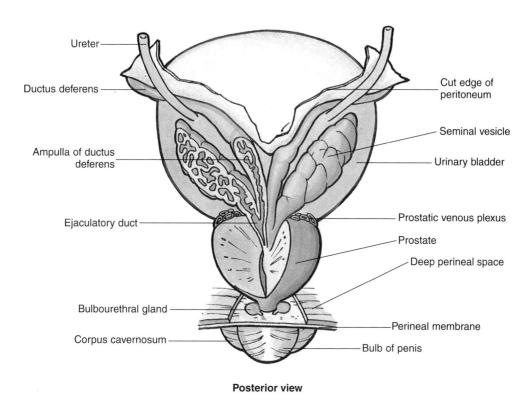

Ureter

Ductus deferens

Ampulla of ductus deferens

Ejaculatory duct

Bulbourethral gland

Corpus cavernosum

Cut edge of peritoneum

Seminal vesicle

Urinary bladder

Prostatic venous plexus

Prostate

Deep perineal space

Perineal membrane

Bulb of penis

Posterior view

FIGURE 3.13 Bladder, prostate and seminal vesicles viewed from the posterior aspect, left side partially dissected.

erens runs along the posterior aspect of the bladder, and expands to form the ampulla before becoming the ejaculatory duct. Just lateral to the ampullae lie the seminal vesicles. The ducts of the seminal vesicles traverse the prostate gland posteriorly and join the vas to form the ejaculatory ducts (Fig. 3.13). Look for the openings of the ejaculatory ducts in the prostatic urethra. As the urethra leaves the prostate gland, it passes through the deep perineal space (sphincter muscle) (membranous urethra) and enters the bulb of the penis to become the **spongy urethra.** G3.16-3.17/C442, 446/ R318-320/ N338, 343, 357-359

5. **Arteries.** The **internal iliac (hypogastric) artery** gives rise to visceral branches, which include the **umbilical, vesical,** and **middle rectals** (note that these may also supply the prostate and seminal vesicles) (Fig. 3.14). Clean and follow these branches to the anterior abdominal wall, bladder, and rectum, respectively. Feel free to remove veins which may obscure your field of dissection. Parietal branches (to skeletal muscles) include the iliolumbar, lateral sacral, obturator, internal pudendal, and superior and inferior gluteal arteries. The **superior gluteal, inferior gluteal,** and **internal pudendal** usually

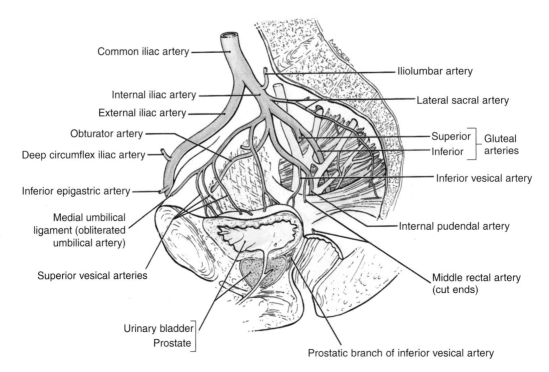

Common iliac artery

Internal iliac artery

External iliac artery

Obturator artery

Deep circumflex iliac artery

Inferior epigastric artery

Medial umbilical
ligament (obliterated
umbilical artery)

Superior vesical arteries

Urinary bladder
Prostate

Iliolumbar artery

Lateral sacral artery

Superior ⎤ Gluteal
Inferior ⎦ arteries

Inferior vesical artery

Internal pudendal artery

Middle rectal artery
(cut ends)

Prostatic branch of inferior vesical artery

FIGURE 3.14 Iliac arteries and branches in the male.

arise in that order from the main portion of the internal iliac artery. Also clean these three branches and note that they leave the pelvis via the greater sciatic foramen. Identify the **obturator artery** leaving the pelvis with the obturator nerve to enter the medial thigh. **A3 2:27:16-2:30:39/ G3.20, 3.32/C410-412/ R324-325/N373-374**

6. Pelvic Diaphragm. Push the pelvic viscera medially and clean the muscles of the pelvic wall and floor. Find the branches of the 2nd, 3rd, and 4th sacral nerves and the **levator ani muscle** (Fig. 3.11). Posteriorly, find the **coccygeus muscle** (often blends with the sacrospinous ligament) inserting into the coccyx and lower sacrum. Looking beneath sacral nerve segments S2-4, find the **piriformis muscle** arising from the sacrum. Its muscle fibers converge and pass through the greater sciatic foramen. **A3 2:15:58-2:27:15/G3.21-3.22/C418/R328/ N335-336**

7. In both sexes, review relevant cross sections of the pelvis. **G3.51/C416, 458/R322, 345/N523**

Now study the female pelvis.

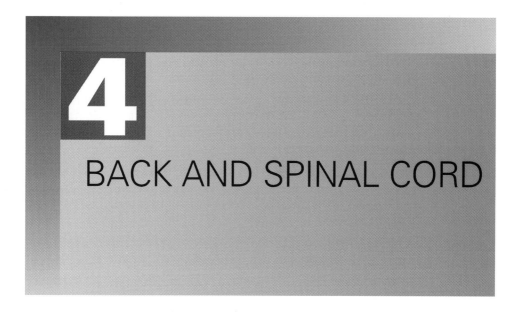

4

BACK AND SPINAL CORD

I. BACK

Learning Objectives

- Throughout Chapter 4, identify structures in bold print unless instructed to do otherwise.
- Identify the diagnostic features of the different vertebrae, the scapula, and the ribs.
- Divide the back muscles into their three functional groups, listing which muscles belong in which group.
- Deduce the attachments (origin and insertion) of back muscles by listing their primary function and understanding how they move the skeletal components.
- Define the term "neurovascular bundle."

Key Concepts

- Vertebrate segmentation
- Functional groups of back muscles
- Neurovascular bundles
- Rib articulation with thoracic vertebrae

A. **Vertebral Column.** The overall body plan of humans is one of segmentation and the vertebral column and spinal nerves offer the best illustration of this important concept.

The vertebral column consists of 33 vertebrae: 7 cervical, 12 thoracic, 5 lumbar, 5 fused sacral (sacrum), and 4 coccygeal vertebrae (coccyx). **A3. 1:35-21:42/ G4.1/C655-657/ R182/ N142**

1. Refer to a skeleton, your cadaver, and each other and note the following bony and surface landmarks:

a. **Scapula.** Identify the **spine, acromion** (highest point of the shoulder, just as the Acropolis is on a hill in Athens), **superior and inferior angle,** and medial or **vertebral border.** **G6.1/C92-93/ R348-349/N393**

b. **Iliac crest.** The **posterior superior iliac spine** where the overlying skin often shows a dimple (especially in females) since no fleshy muscle lies beneath this point.   **G4.28/ C624, 628/ R209/ N160**

c. **Occipital bone** (skull) and the **external occipital protuberance (inion)** and the **nuchal lines.**   **G7.2/ C757/ R33/ N5**

d. **Mastoid process** on the skull.   **G7.2/ C757/ R25/ A7.9/ N2**

e. **Cervical vertebrae.** Identify the **transverse process** and its **transverse foramen.** The **spinous processes** are bifid and the spine of C7 (not bifid) is the most prominent. Palpate your own C7 spinous process (vertebrae prominens). **G4.9/ C639-644/ R185, 191/N12-13**

f. **Atlas** (C1). Lacks a body and spinous process. The **axis** is C2 and has a bifid spine.

g. **Vertebral foramen.** Successive foramina form the **vertebral canal** which contains the spinal cord and its membranes (meninges).

h. **Thoracic and lumbar vertebrae.** Note the weight-bearing **body,** and the vertebral arch formed by two **pedicles** (roots), and two **laminae.** A **transverse process** projects laterally on each side and has **facets** for rib articulation. Identify the **superior and inferior articular processes** and the **spinous process.**   **G4.12/ C658-662/ R184-185/N143-144**

i. **Rib** and its **head.** Articulates with two vertebral bodies and the intervening disc. The **tubercle** articulates with the transverse process of the vertebra with the same segmental number while the head articulates both with the vertebra above and at the same segmental level (e.g., head of rib 5 articulates with vertebral bodies T4 and T5, and the tubercle with the transverse process of T5).   **G1.9/ C658-664/ R189/ N171-172**

j. Fibrocartilaginous **intervertebral discs** unite adjacent vertebral bodies.   **G4.24/ C665-667/ R190/N144**

k. **Intervertebral foramen** (hole). Transmits the spinal nerve.

B. Back Muscles. The back muscles are divided functionally into three groups: **superficial** (attaches upper limb to the axial skeleton), **intermediate** (for respiration), and **deep** ("intrinsic" or true back muscles for movements of the spine, innervated by dorsal primary rami of spinal nerves) (Table 4.1).

TABLE 4.1
INTRINSIC BACK MUSCLES

Muscles	Origin	Insertion	Nerve Supply[a]	Main Actions
Superficial Layer				
Splenius	Arises from ligamentum nuchae and spinous processes of C7–T3 or T4 vertebrae	*Splenius capitis:* fibers run superolaterally to mastoid process of temporal bone and lateral third of superior nuchal line of occipital bone *Splenius cervicis:* posterior tubercles of transverse processes of C1–C3 or C4 vertebrae		*Acting alone,* they laterally bend and rotate head to side of active muscles; *acting together,* they extend cervical and upper thoracic vertebrae
Intermediate Layer				
Erector spinae	Arises by a broad tendon from posterior part of iliac crest, posterior surface of sacrum, sacral and inferior lumbar spinous processes, and supraspinous ligament	*Iliocostalis:* lumborum, thoracis and cervicis; fibers run superiorly to angles of lower ribs and cervical transverse processes *Longissimus:* thoracis, cervicis, and capitis; fibers run superiorly to ribs between tubercles and angles, to transverse processes in thoracic and cervical regions, and to mastoid process of temporal bone *Spinalis:* thoracis, cervicis, and capitis; fibers run superiorly to spinous processes in the upper thoracic region and to skull	Dorsal rami of spinal nn.	*Acting bilaterally,* they extend vertebral column; as back is flexed they control movement by gradually lengthening their fibers; *acting unilaterally,* they laterally bend vertebral column
Deep Layer				
Transversospinal	Semispinalis arises from thoracic and cervical transverse processes	*Semispinalis:* thoracis, cervicis, and capitis; fibers run superomedially and attach to occipital bone and spinous processes in thoracic and cervical regions, spanning four to six segments		Extend head and thoracic and cervical regions of vertebral column and rotate them contralaterally
	Multifidus arises from sacrum and ilium, transverse processes of T1–T3, and articular processes of C4–C7	Fibers pass superomedially to spinous processes, spanning two to four segments		Stabilizes vertebrae during local movements of vertebral column
	Rotatores arise from transverse processes of vertebrae; best developed in thoracic region	Pass superomedially and attach to junction of lamina and transverse process of vertebra of origin or into spinous process above their origin, spanning one to two segments		Stabilize vertebrae and assist with local extension and rotary movements of vertebral column

[a]Most back muscles are innervated by dorsal rami of spinal nerves, but a few are innervated by ventral rami. Anterior intertransversarii of cervical region are supplied by ventral rami.

			Nerve	Main
TABLE 4.1 (Continued)				
INTRINSIC BACK MUSCLES				
Muscles	Origin	Insertion	Supply[a]	Actions
Minor Deep Layer				
Interspinales	Superior surfaces of spinous processes of cervical and lumbar vertebrae	Inferior surfaces of spinous processes of vertebrae superior to vertebrae of origin	Dorsal rami of spinal nn.	Aid in extension and rotation of vertebral column
Intertransversarii	Transverse processes of cervical and lumbar vertebrae	Transverse processes of adjacent vertebrae	Dorsal and ventral rami of spinal nn.	Aid in lateral bending of vertebral column; acting bilaterally, they stabilize vertebral column
Levatores costarum	Tips of transverse processes of C7 and T1–T11 vertebrae	Pass inferolaterally and insert on rib between its tubercle and angle	Dorsal rami of C8–T11 spinal nn.[b]	Elevate ribs, assisting inspiration; assist with lateral bending of vertebral column

[b]Levatores costarum were once said to be innervated by ventral rami, but investigators now agree that they are innervated by dorsal rami.

Make the skin incisions shown in Figure 4.1. First, make a midline incision along the dashed line A-B-C-D. Then, make horizontal incisions A to E, B to F, C to G, and D to H and remove the skin and tela subcutanea (superficial fascia) down to the level of the underlying muscles. Reflect the skin flaps resulting from your horizontal cuts as far laterally as the midaxillary line. As you skin the back, notice small nerves and vessels piercing the muscles and coursing upward into the skin. These are **neurovascular bundles** and the nerves that innervate the deep intrinsic back muscles are dorsal branches of the dorsal primary rami. **G4.41-4.43/ C628, 634/ R212/N163**

1. Clean the **trapezius** and **latissimus dorsi muscles** (Fig. 4.2). At a point about 3 cm inferolateral to the inion, try to find the **greater occipital nerve** and just lateral to it the **occipital artery.** Because of the dense nature of the deep fascia in this area you may have difficulty finding these structures. Observe two triangles bounded by the latissimus dorsi, the triangle of auscultation and the lumbar triangle.

   The triangle of auscultation is free of overlying muscles while the lumbar triangle can be the site of "lumbar hernias." **G4.29/ C628-629/ R212/ N160**

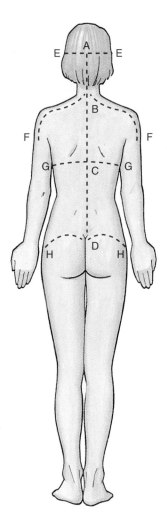

FIGURE 4.1. Skin incisions on the back.

2. Trapezius. Study the extent of the trapezius (Fig. 4.2) and then reflect the muscle as follows: retract the shoulder posteriorly to relax the trapezius and separate the muscle from underlying structures by passing your hand or scissors beneath the muscle. Cut the muscle very close to the spinous processes inferiorly all the way to the inion superiorly, and cut its attachment to the scapular spine and acromion, *leaving the trapezius attached only to the clavicle.* Reflect the muscle laterally.   **G4.29/ C629/ R209, 220/N160**

On the deep surface of the trapezius find its innervation, the **spinal accessory nerve** (11th cranial nerve).   **G4.29/ C634/ R221/N163**

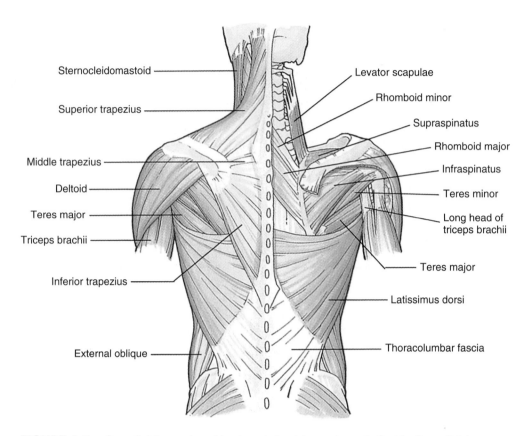

FIGURE 4.2. Superficial muscles of back and shoulder regions. *Left side,* Superficial dissection. *Right side,* Deeper dissection.

3. **Latissimus dorsi.** This muscle arises from the vast **thoracolumbar fascia** (Fig. 4.2). Superiorly, the fibers of this muscle converge to form a tendon that inserts into the humerus. Lift this muscle by sliding your fingers beneath it, and then carefully cut through the thoracolumbar fascia and reflect the muscle laterally. **G4.29/ C628-629/ R212/ N160**

4. Next, study the three remaining muscles of the superficial group: the **rhomboids** (major and minor) and the **levator scapulae** (Fig. 4.2). Detach the rhomboids from the spinous processes and reflect them laterally. Do not reflect the levator scapulae. **G4.30/ C629/ R212, 221/N160, 163**

5. In your textbook, study the primary action of these muscles. Focus your attention on the "actions" of muscles rather than on their origins and insertions. If you understand the action of a muscle, you can usually deduce its attachments. If your instructor emphasizes origins and insertions, refer to Table 4.1 and your atlas and textbook for descriptions.

II. DEEPER BACK MUSCLES AND SPINAL CORD

Learning Objectives

- List the actions of the intermediate and deep (intrinsic) back muscles.
- Explain how the intrinsic back muscles receive their innervation by dorsal primary rami of spinal nerves.
- Identify the features of the spinal cord and its meningeal coverings.
- Draw a typical spinal nerve showing its efferent and afferent components and how it originates from the spinal cord.
- Define the term "dermatome."

Key Concepts

- Dermatome
- Region of back and muscles innervated by dorsal primary rami
- Meninges
- Vertebral venous plexus

A. Intermediate Muscles. Observe the **serratus posterior superior** and **serratus posterior inferior,** and then cut both muscles at their origin from the vertebrae and reflect them laterally. These muscles attach to ribs and are respiratory in function.   **G4.30/ C630/ R209, 221/ N161**

B. Deep Muscles. Deep muscles include the splenius capitis and cervicis, semispinalis capitis, erector spinae, and transversospinalis group (Table 4.1) (Fig. 4.3). Deep to the semispinalis capitis lie the suboccipital muscles.

1. Splenius capitis and cervicis. On one side of the body only, cut the splenius from the **ligamentum nuchae** and spinous processes and reflect it to expose the **semispinalis capitis.** **G4.31/ C630/ R223-224/N161**

2. Erector spinae. Distinguish the three vertically running columns of the erector spinae: **iliocostalis, longissimus,** and **spinalis** (Fig. 4.3). Acting together, these muscles extend the vertebral column. Acting unilaterally, the three muscle columns bend the vertebral column laterally.   **A3. 21:43-26:40/ G4.31/ C630-631/ R210/ N161**

3. Transversospinalis. These short muscles occupy the groove between the transverse and spinous processes and include the semispinalis, multifidus, and rotatores muscles (Fig. 4.3). Note

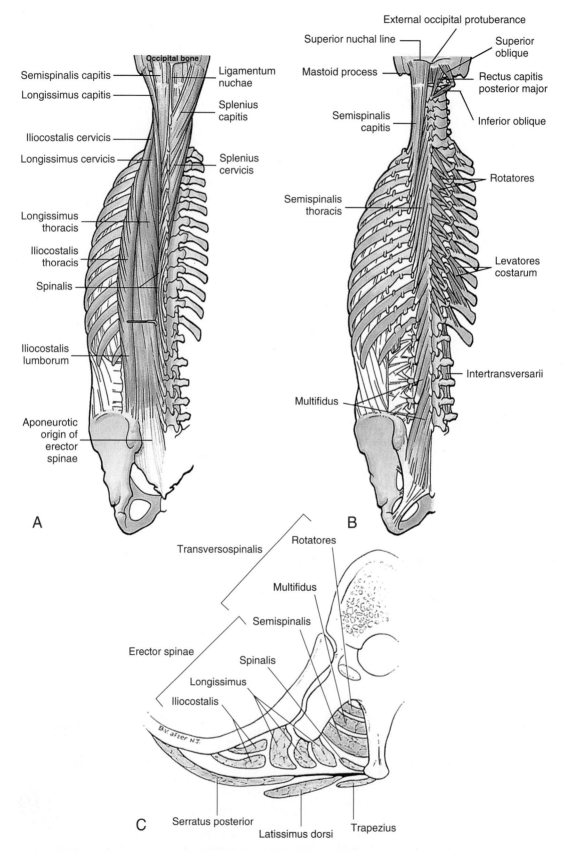

FIGURE 4.3. Intrinsic back muscles. A. Erector spinae, splenius, and semispinalis. B. Dissection of the back showing transversospinalis. C. Partial transverse section of back showing muscles.

these muscles in your atlas but there is no need to dissect them unless instructed to do so (clinicians often call these muscles **"paravertebral"** muscles).   G4.32-4.33/ C632-633/ R211/ N162

C. **Vertebral Canal and Spinal Cord.** Place a wooden block under the pelvis to reduce the concavity of the lumbar region and then remove all the dorsal paravertebral musculature from T6 to L5 with a scalpel. Note that **supraspinous** and **interspinous ligaments** attach adjacent spines to each other.   A3. 26:41-34:13/ G4.26/ C663-667/ R190/ N146

Remove the lamina (laminectomy) about 2 cm lateral to the midline using a chisel (directed at a 45 degree angle) and mallet. Use a bone pliers to remove any remaining bone (be careful to protect your eyes from flying bone chips) and open the vertebral canal from T6 to L5. Observe the **ligamenta flava,** strong elastic ligaments that connect adjacent laminae.   G4.26/ C667/ R190/ N146

1. **Spinal membranes.** Identify the **epidural (extradural) space,** which contains fat and the vertebral venous (Batson's) plexus. If drained of venous blood, this plexus may not be visible in your cadaver. However, read about this venous plexus in your textbook as these veins can provide a route for metastasis of cancer cells to the vertebrae and brain.   G4.27/ C683-689/ R218/ N156, 159

Incise the **dura mater** along the dorsal midline. Identify the delicate underlying avascular **arachnoid mater.** Incise the arachnoid mater to open the subarachnoid space. This space normally contains the cerebrospinal fluid (but not in the cadaver because it has been absorbed into the tissues).   G4.43-4.45/ C684-688/ R216, 218/ N155

2. Observe the following (Fig.4.4):   A3. 26:41-34:13/ G4.41-4.46/ C681-690/ R216-218/ N148-149, 155-156

  a. **Spinal cord.** Surrounded by the **pia mater,** much like a nylon stocking surrounds a leg.

  b. **Denticulate ligaments.** Usually 20-21 pairs of white colored pia mater extensions (toothlike in appearance) that anchor the spinal cord laterally to the dura.

  c. **Ventral and dorsal roots.** Follow them as they pierce the dura and enter the intervertebral foramen.

  d. **Dorsal root ganglion (DRG).** In the thoracic region, expose one or two DRGs in the intervertebral foramina and see if you can identify a **spinal nerve** and its ventral and dorsal primary rami. DRG contain the nerve cell bodies of sensory nerves.

  e. **Conus medullaris.** End of the spinal cord proper, between the L1 and L2 vertebra. At birth, the conus lies at the level of L3 or L4.

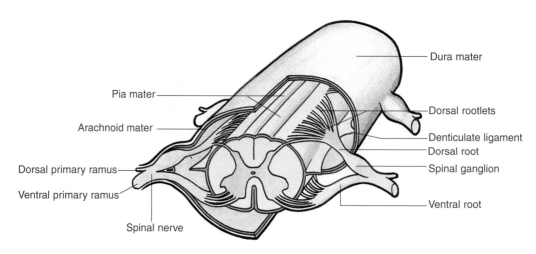

FIGURE 4.4.   Spinal cord and meninges.

    f. Cauda equina. Collection of ventral and dorsal roots that
    look like a "horse's tail."

    g. Filum terminale. A fine strand of pia mater (usually whiter
    in color than the nerve roots) which extends to the end of
    the dural sac at S2 (also the end of the subarachnoid space),
    then pierces the dura and, with a dural covering, passes

FIGURE 4.5.   Simplified transverse section of thorax. Nerves shown on right and arteries on
left.

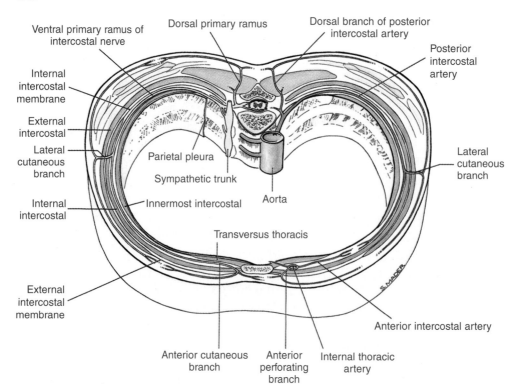

through the sacral hiatus to anchor the spinal cord to the coccyx.   **G4.46/ C681-682/ R217/ N148**

3. Note that there are 31 pairs of spinal nerves (8 cervical, 12 thoracic, 5 lumbar, 5 sacral, and 1 coccygeal pair).

4. In your textbook and atlas, review the distribution of a spinal nerve (Fig. 4.5), understand what a dermatome is, and read about the somatic and autonomic portions of the peripheral nervous system.   **G4.41-4.51/ C625, 681-690/ R204, 214/ N150, 153-154, 156**

D. Suboccipital Region. This dissection may be optional. Please check with your instructor. Regardless, you should view an atlas illustration and appreciate that this region between the occipital bone and the upper two cervical vertebrae contains muscles that maintain the posture of the neck and move the head.

1. Detach the semispinalis capitis muscle bilaterally near the occipital bone and reflect it inferiorly. Preserve the **greater occipital nerve** (dorsal ramus of C2).

2. Identify the **inferior oblique** (obliquus capitis inferior), **superior oblique** (obliquus capitis superior), and **rectus major** (rectus capitis posterior major) muscles bounding the suboccipital triangle (Fig. 4.6). The **rectus minor** lies medial to the major.   **G4.36-4.37/ C635-638/ R210-211/ N164**

FIGURE 4.6.   Suboccipital region.

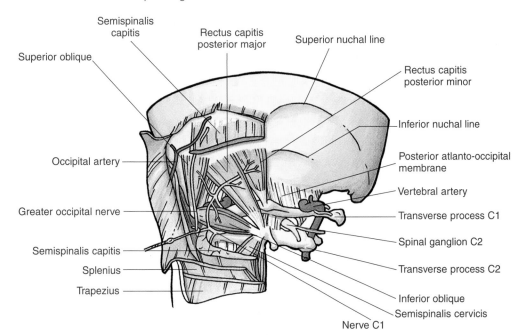

3. Find the **suboccipital nerve** (dorsal ramus of C1, which contains largely motor nerves to the muscles of the triangle) and the **vertebral artery,** which lies deep and can be seen passing out of the transverse foramen of the atlas before entering the foramen magnum to supply blood to the brain.

4. The suboccipital muscles extend the head on the atlas (atlantooccipital joint) and rotate the head on the atlas and axis (atlantoaxial joint).

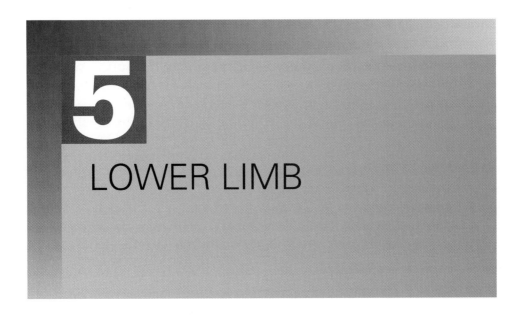

LOWER LIMB

I. ANTERIOR AND MEDIAL THIGH

Learning Objectives

- Throughout Chapter 5, identify structures in bold print unless instructed to do otherwise.
- Identify the key superficial veins of the lower limb.
- Illustrate the dermatome pattern of the lower limb and list the major cutaneous nerves of the thigh.
- Identify the main bony features of the pelvis and femur.
- List the contents of the femoral triangle from lateral to medial.
- On a diagram, be able to label the key branches of the femoral artery.
- Describe the function and innervation of the muscles of the anterior and medial thigh.

Key Concepts

- Limb rotation during development
- Femoral triangle
- Muscular compartments
- Deep and superficial venous return
- Collateral circulation around major joints

A. Introduction. The lower limb is divided into the thigh (region between the hip and knee), the leg (region between the knee and ankle), and the foot. The flexor muscle compartments of the lower limb generally are posterior while the extensor muscle compartments are generally anterior. Realize that this is about 180 degrees different from the upper limb (where flexor compartments are anterior) due to the embryologic rotation of the upper and lower limbs.

When dissecting the lower limb, focus on the muscles as groups within a fascial compartment, noting their actions, innervation, and blood supply. In other words, use the tables provided to summarize information. For those courses that do emphasize origins and insertions of the muscles, the tables provide this information. Check with your instructor before memorizing these details, however.

Make the skin incisions indicated in Figure 5.1. Remove both the skin and the tela (superficial fascia) so that your dissection comes down onto the deep fascia investing the muscles. (Some instructors may prefer that the skin be removed in vertical strips from proximal to distal "with the grain" in terms of the vertically oriented veins and cutaneous nerves. Please check with your instructor). Note that the deep fascia is strong and envelops the muscles like a nylon stocking. As you remove the skin, note the location and identify the **great** and **small (lesser) saphenous** (Gr. manifest; visible) **veins.** The **saphenous nerve** courses with the great saphenous vein and the **sural nerve** courses with the small saphenous vein. Study the origin of these veins in your atlas and note the cutaneous nerves of the lower limb. The superficial veins of the limb are connected to deeper lying veins by perforating veins that pierce the deep fascia.   **G5.4-5.5/ C486-487, 538/ R444/ N508-509**

FIGURE 5.1.   Dashed lines show skin incisions on the lower limb.

Anterior view

B. Femoral Triangle and Sheath

1. Identify the **femoral triangle** and its boundaries, the **inguinal ligament, sartorius muscle,** and **adductor longus muscle** (Fig. 5.2).

2. Follow the **great saphenous vein** cranially until it disappears in the deep fascia (**saphenous opening** or fossa ovalis) and enters the femoral vein. Slit open a portion of the great saphenous vein to observe its valves (10 to 20 such valves exist along its course).   **G5.11-5.12/C489-491/R451/ N466**

3. Find **superficial inguinal lymph nodes** medial to the saphenous opening. These superficial nodes drain into external iliac nodes or **deep inguinal nodes** that lie medial and deep to the femoral vein.

4. The **femoral sheath** envelops the **femoral artery, vein,** and medially situated lymph nodes and vessels (spells NAVEL from lateral to medial, if one includes the femoral nerve which lies lateral to but outside the femoral sheath (Fig. 5.2). The "E" signifies an "empty" space between the femoral vein and medial-lying lymph nodes). The **femoral canal** is the medial portion of the sheath and contains lymphatics and connective tissue.   **A2.31:10-39:10/ G5.12/ C495/ R451/ N466**

C. Anterior Thigh

1. First, study the following bony landmarks on a skeleton: **A2. 00:49-11:08/ G5.1/ C376, 560, 569/ R411-416/ N453-455, 478**

    a. Anterior superior iliac spine. (The inguinal ligament attaches here)

FIGURE 5.2. Femoral triangle and fascia.

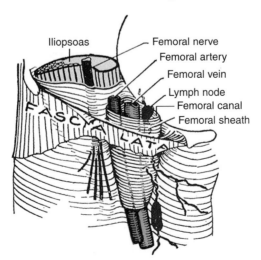

   b. **Anterior inferior iliac spine.**

   c. **Pubic tubercle.** (The inguinal ligament also attaches here)

   d. **Greater trochanter of femur.**

   e. **Lesser trochanter of femur.**

   f. **Lateral condyle and epicondyle of femur.**

   g. **Medial condyle and epicondyle of femur.**

   h. **Adductor tubercle.**

   i. **Linea aspera.** Attachment of adductor magnus muscle, and several other muscles.

   j. **Patella.**

   k. **Tibial tuberosity of tibia.**

2. Clean the **femoral artery** and **vein** in the femoral triangle and the **profunda** (L., deep) **femoris artery.** Also clean the **lateral** and **medial femoral circumflex arteries.** The medial circumflex branch supplies most of the blood to the head and neck of the femur. Remove smaller, deeper veins if they obscure your dissection. Identify the large **femoral nerve** *lateral* to the femoral artery and note its muscular branches to the sartorius and rectus femoris muscles (Fig. 5.3).   **G5.21/ C498/ R442/N466-467**

3. Cut the **sartorius muscle** several centimeters below its proximal attachment and reflect it inferiorly to expose the **adductor (Hunter's) canal.** The femoral vessels and **saphenous nerve** pass through this canal but the saphenous nerve continues distally by passing between the sartorius and gracilis superficially; note that the femoral vein lies *posterior* to the artery in the canal. The femoral vessels disappear by passing through the **adductor hiatus** at the distal end of the canal to assume their position behind the knee (now they become popliteal vessels).   **G5.21/ C498/ R452/ N466-467**

4. Examine the **deep fascia (fascia lata;** L., broad) of the thigh. Laterally, this dense fascia forms the **iliotibial tract.** Superiorly, this fascia envelops the **tensor fasciae latae muscle.** Split open the deep fascia between the rectus femoris and vastus lateralis and free the iliotibial tract from the underlying vastus lateralis.   **G5.23/ C492, 503/ R456/N460**

5. Clean and identify the **quadriceps femoris (vastus lateralis, medialis** and **intermedius,** and the **rectus femoris) muscle** (Fig. 5.3). The tendons of these muscles unite to form the **patellar tendon** (ligament) which inserts on the tibial tuberosity. Note that the sartorius and rectus femoris muscles cross two joints. Study the actions of these thigh muscles (Table 5.1).   **A2. 31:10-39:10, 58:20-1:07:44/ G5.17/ C493/ R452-453/ N467**

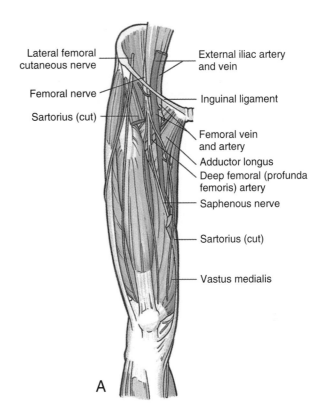

Lateral femoral cutaneous nerve

Femoral nerve

Sartorius (cut)

External iliac artery and vein

Inguinal ligament

Femoral vein and artery

Adductor longus

Deep femoral (profunda femoris) artery

Saphenous nerve

Sartorius (cut)

Vastus medialis

A

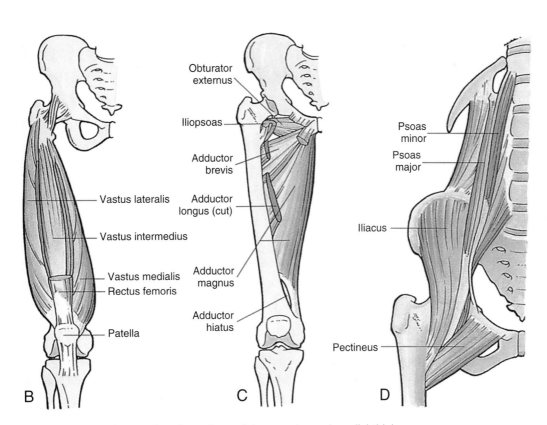

Vastus lateralis

Vastus intermedius

Vastus medialis
Rectus femoris

Patella

B

Obturator externus

Iliopsoas

Adductor brevis

Adductor longus (cut)

Adductor magnus

Adductor hiatus

C

Psoas minor

Psoas major

Iliacus

Pectineus

D

FIGURE 5.3. Successive dissections of the anterior and medial thigh.

TABLE 5.1
ANTERIOR THIGH MUSCLES

Muscle	Proximal Attachment	Distal Attachment	Innervation[a]	Main Actions
Iliopsoas				
Psoas major	Sides of T12–L5 vertebrae and discs between them; transverse processes of all lumbar vertebrae	Lesser trochanter of femur	Ventral rami of lumbar nerves (**L1, L2,** and L3)	Acting jointly in flexing the hip joint and in stabilizing this joint[b]
Iliacus	Iliac crest, iliac fossa, ala of sacrum, and anterior sacroiliac ligaments	Tendon of psoas major, lesser trochanter, and femur distal to it	Femoral nerve (**L2** and L3)	
Tensor fasciae latae	Anterior superior iliac spine and anterior part of iliac crest	Iliotibial tract that attaches to lateral condyle of tibia	Superior gluteal (**L4** and **L5**)	Abducts, medially rotates, and flexes hip; helps to keep knee extended; steadies trunk on thigh
Sartorius	Anterior superior iliac spine and superior part of notch inferior to it	Superior part of medial surface of tibia	Femoral nerve (L2 and L3)	Flexes, abducts, and laterally rotates the hip joint; flexes the knee joint[c]
Quadriceps femoris Rectus femoris	Anterior inferior iliac spine and ilium superior to acetabulum			Extends the knee joint; rectus femoris also steadies hip joint and helps iliopsoas to flex the hip
Vastus lateralis	Greater trochanter and lateral lip of linea aspera of femur	Base of patella and by patellar ligament to tibial tuberosity	Femoral nerve (L2, **L3,** and **L4**)	
Vastus medialis	Intertrochanteric line and medial hip of linea aspera of femur			
Vastus intermedius	Anterior and lateral surfaces of body of femur			

[a]Numbers indicate spinal cord segmental innervation of nerves [e.g., **L1, L2,** and L3 indicate that nerves supplying psoas major are derived from first three lumbar segments of the spinal cord: boldface type (**L1, L2**) indicates main segmental innervation]. Damage to one or more of these spinal cord segments or to motor nerve roots arising from them results in paralysis of the muscles concerned.

[b]Psoas major is also a postural muscle that helps control deviation of trunk and is active during standing.

[c]Four actions of sartorius (L, sartor, tailor) produce the once common crosslegged sitting position used by tailors—hence the name.

D. Medial Thigh

1. Remove the deep fascia of the medial thigh and study the adductor muscles. Identify the **pectineus, adductor longus,** and **gracilis muscles** (Fig. 5.3). Distally, the gracilis inserts on the tibia while the other two muscles insert into the linea aspera of the femur. The profunda femoral vessels pass between the

pectineus and adductor longus. **G5.19/C499-500/R452-453/ N458-459, 466-467**

2. Cut the **adductor longus** several centimeters below its proximal attachment and reflect it to reveal the **adductor brevis.** Identify branches of the anterior division of the **obturator nerve** innervating these adductor muscles. **G5.19/ C501/ R453/ N459, 467, 503**

3. Separate the adductor brevis and **adductor magnus** to view posterior division of the obturator nerve passing between these two muscles. Cut the brevis close to its proximal attachment and view the magnus. Identify the **adductor hiatus** in the muscle tendon. Now cut the pectineus close to its origin and try to identify the **obturator externus,** which covers the obturator membrane. Study the actions of the adductor muscles and realize that the gracilis crosses two joints (Table 5.2). **G5.21/ C501/ R429, 453/N467, 471**

TABLE 5.2
MEDIAL THIGH MUSCLES

Muscle	Proximal Attachment	Distal Attachment	Innervation[a]	Main Actions
Pectineus	Superior ramus of pubis	Pectineal line of femur, just inferior to lesser trochanter	Femoral nerve (**L2** and L3); may receive a branch from obturator nerve	Adducts and flexes hip; assists with medial rotation of thigh
Adductor longus	Body of pubis inferior to pubic crest	Middle third of linea aspera of femur	Obturator nerve (L2, **L3,** and L4)	Adducts hip
Adductor brevis	Body and inferior ramus of pubis	Pectineal line and proximal part of linea aspera of femur	Obturator nerve (L2, **L3,** and L4)	Adducts hip and to some extent flexes it
Adductor magnus	Inferior ramus of pubis, ramus of ischium (adductor part), and ischial tuberosity (hamstring part)	Gluteal tuberosity, linea aspera, medial supracondylar line (adductor part), and adductor tubercle of femur (hamstring part)	*Adductor part:* obturator nerve (L2, **L3,** and **L4**) *Hamstring part:* tibial part of sciatic nerve (**L4**)	Adducts hip; its adductor part also flexes hip, and its hamstring part extends it
Gracillis	Body and inferior ramus of pubis	Superior part of medial surface of tibia	Obturator nerve (**L2** and L3)	Adducts hip, flexes knee, and helps rotate it medially
Obturator externus	Margins of obturator foramen and obturator membrane	Trochanteric fossa of femur	Obturator nerve (L3 and **L4**)	Laterally rotates hip; steadies head of femur in acetabulum

Collectively, the first five muscles listed are the adductors of the hip, but their actions are more complex (e.g., they act as flexors of the hip joint during flexion of the knee joint and are active during walking).

[a]See Table 5.1 for explanation of segmental innervation.

II. GLUTEAL REGION AND POSTERIOR THIGH

Learning Objectives

- List the actions of the gluteal muscles on the hip.
- List the actions of the muscles of the posterior thigh (hip and knee).
- Describe the distribution of the gluteal nerves and sciatic nerve, and know what nerve roots contribute to the formation of the sciatic nerve.
- On a cross section of the thigh, identify the functional muscle compartments, the blood supply to the compartment, and the nerve innervating the muscles of the compartment.

Key Concepts

- Muscles that cross a joint usually act on that joint
- Importance of bursae
- Site for gluteal intramuscular injections

A. Bony Features

1. Identify the following landmarks on a skeleton: A2. 00:49-11:08/ G3.1, 5.1/ C392,560,566/ R420-421/ A5.41/ N453, 455

a. **Greater sciatic notch.**

b. **Lesser sciatic notch.**

c. **Ischial tuberosity.**

d. **Ischial spine.**

e. **Sacrotuberous ligament** (on model).

f. **Sacrospinous ligament** (on model).

g. **Greater and lesser sciatic foramen** (on model).

h. **Greater trochanter of femur.**

i. **Intertrochanteric crest.**

j. **Gluteal tuberosity.** For the attachment of the gluteus maximus muscle.

B. Gluteus Maximus

1. Make the skin incisions shown in Figure 5.4A and reflect the skin flaps. Clean the **gluteus maximus muscle** (a powerful extensor of the hip), define its borders, and note that some of the **gluteus medius** is visible superior to the maximus (Fig. 5.4B). Push your fingers into the space between the medius and maximus to separate these two muscles and then cut the max-

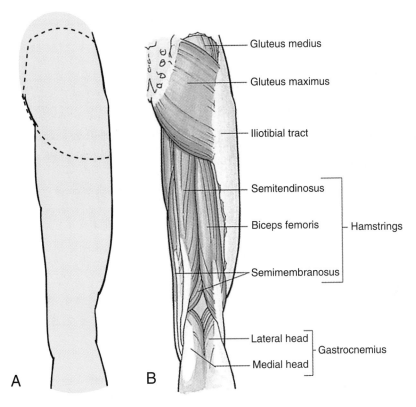

FIGURE 5.4.   A. Dashed lines show skin incisions to expose the gluteal region. B. Superficial dissection of the gluteus maximus and hamstring muscles.

imus close to the posterior surfaces of the ilium, sacrum, and coccyx. If the pelvis and perineum were dissected previously, the maximus may already be reflected. Detach it from the **sacrotuberous ligament** and reflect it laterally toward the greater trochanter. Cut any neurovascular bundles (**inferior gluteal vessels** and **nerves**) attached to the muscle's under side. A large bursa lies between the muscle and the greater trochanter, the **trochanteric bursa** (usually collapsed in the cadaver). This bursa protects the muscle as it slides over the bone.   **G5.29/ C506/ R454-455/N461**

C. Deep Structures

1. Identify the sciatic nerve as it enters the gluteal region inferior to the **piriformis** (L., pear-shaped) **muscle** (Fig. 5.5). Also inferior to the piriformis, identify the **inferior gluteal vessels** and **nerves, pudendal nerve,** and **internal pudendal vessels.** The pudendal neurovascular bundle passes around the **sacrospinous ligament** and enters the lesser sciatic foramen to travel to the perineum (via the pudendal canal or Alcock's canal). Realize that because of the presence of neurovascular bundles

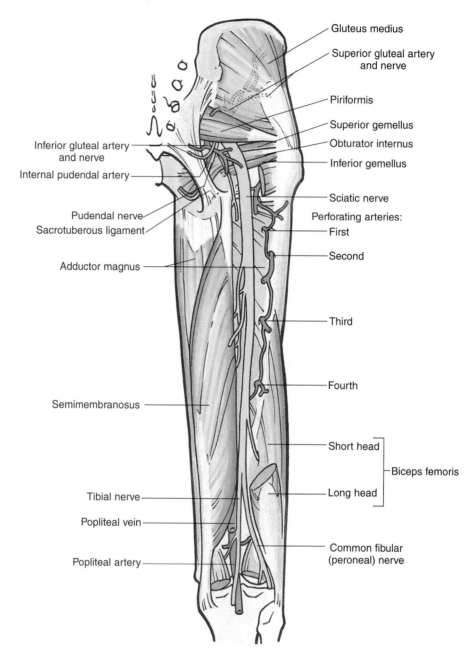

FIGURE 5.5.   Gluteal region and posterior thigh.

medially, intragluteal injections are given in the superolateral region.   **A2. 11:08-45:33/ G5.28/ C506/ R455, 457/ N468-469**

2. Identify the **obturator internus** (comes through the lesser sciatic foramen) and the two **gemelli** (twin) **muscles** (superior and inferior). Inferior to the inferior gemellus, find the **quadratus femoris**. Try to find the **obturator externus tendon**. **G5.29/ C511/ R457/ N469**

3. Superior to the piriformis, identify the **superior gluteal vessels** and **nerves,** largely covered by the gluteus medius. Separate the **gluteus medius** and **minimus muscles** (often fused). Cut the medius several centimeters above its insertion on the greater trochanter and reflect the muscle superiorly to see the minimus and superior gluteal neurovascular bundle. Try to follow the superior gluteal nerve to the **tensor fasciae latae muscle,** which it also innervates.

Study the actions of the gluteal muscles (Table 5.3). **G5.27-5.28/ C507/ R455/ N469**

TABLE 5.3
MUSCLES OF GLUTEAL REGION

Muscle	Proximal Attachment	Distal Attachment	Innervation[a]	Main Actions
Gluteus maximus	Ilium posterior to posterior gluteal line, dorsal surface of sacrum and coccyx, and sacrotuberous ligament	Most fibers end in iliotibial tract that inserts into lateral condyle of tibia; some fibers insert on gluteal tuberosity of femur	Inferior gluteal nerve (L5, **S1,** and **S2**)	Extends hip and assists in its lateral rotation; steadies thigh and assists in raising trunk from flexed position
Gluteus medius	External surface of ilium between anterior and posterior gluteal lines	Lateral surface of greater trochanter of femur	Superior gluteal nerve (**L5** and S1)	Abduct and medially rotate hip; steady pelvis on leg when opposite leg is raised
Gluteus minimus	External surface of ilium between anterior and inferior gluteal lines	Anterior surface of greater trochanter of femur		
Piriformis	Anterior surface of sacrum and sacrotuberous ligament	Superior border of greater trochanter of femur	Branches of ventral rami of S1 and S2	Laterally rotate extended hip and abduct flexed hip; steady femoral head in acetabulum
Obturator internus	Pelvic surface of obturator membrane and surrounding bones	Medial surface of greater trochanter of femur[b]	Nerve to obturator internus (L5 and **S1**); superior gemellus: same nerve supply as obturator internus inferior gemellus: same nerve supply as quadratus femoris	
Gemelli, superior and inferior	Superior, ischial spine; inferior, ischial tuberosity			
Quadratus femoris	Lateral border of ischial tuberosity	Quadrate tubercle on intertrochanteric crest of femur and inferior to it	Nerve to quadratus femoris (L5 and S1)	Laterally rotates hip[c]; steadies femoral head in acetabulum

[a]See Table 5.1 for explanation of segmental innervation.

[b]Gemelli muscles blend with tendon of obturator internus muscle as it attaches to greater trochanter of femur.

[c]There are six lateral rotators of the hip: piriformis, obturator internus, gemelli (superior and inferior), quadratus femoris, and obturator externus. These muscles also stabilize the hip joint.

TABLE 5.4
POSTERIOR THIGH MUSCLE

Muscle	Proximal Attachment	Distal Attachment	Innervation[a]	Main Actions
Semitendinosus	Ischial tuberosity	Medial surface of superior part of tibia Posterior part of medial condyle of tibia	Tibial division of sciatic n. (**L5, S1,** and S2)	Extend hip; flex knee and rotate it medially; when hip and knee are flexed, they can extend trunk
Semimembranosus				
Biceps femoris	*Long head:* ischial tuberosity *Short head:* Linea aspera and lateral supracondylar line of femur	Lateral side of head of fibula; tendon is split at this site by fibular collateral ligament of knee	Long head: tibial division of sciatic n. (L5, **S1,** and S2) Short head: common fibular (peroneal) division of sciatic n. (L5, **S1,** and S2)	Flexes knee and rotates it laterally; extends hip (e.g., when starting to walk)

[a]Collectively these three muscles are known as hamstrings.
[b]See Table 5.1 for explanation of segmental innervation.

D. Posterior Thigh

1. Identify the hamstring muscles, the **semitendinosus, semi-membranosus,** and **long head of the biceps femoris.** These muscles all originate from the ischial tuberosity and cross two joints. Confirm that the **short head of the biceps femoris** arises from the shaft of the femur, crosses only one joint and is *not* part of the hamstring group. Study the attachments of the hamstring muscles around the knee joint and learn their actions (Table 5.4). **G5.25/ C510-511/ R456-457/ N457, 468**

2. Identify the **sciatic nerve** (tibial and peroneal divisions) in the thigh and note branches innervating the hamstring muscles. Realize that the hamstring muscles receive blood from perforating branches of the deep femoral artery, and the medial circumflex femoral artery (Fig. 5.5).

III. POPLITEAL FOSSA AND LEG

Learning Objectives

- Identify the main features of the tibia and fibula, and identify the metatarsals.
- Describe the primary actions of the leg muscles (on knee, ankle, and toes).
- List the arteries responsible for the vascular anastomosis around the knee.
- On a cross section of the leg, identify the functional muscular compartments, the vascular supply to the compartment, and the innervation of the muscles in that compartment.

Key Concepts

- Popliteal fossa
- Collateral circulation around major joints
- Importance of testing tendon reflexes

A. Popliteal Fossa. The popliteal fossa is a diamond-shaped region behind the knee. It is bound by the hamstrings and the two heads of the gastrocnemius muscle (Fig. 5.6).

FIGURE 5.6. Popliteal fossa.

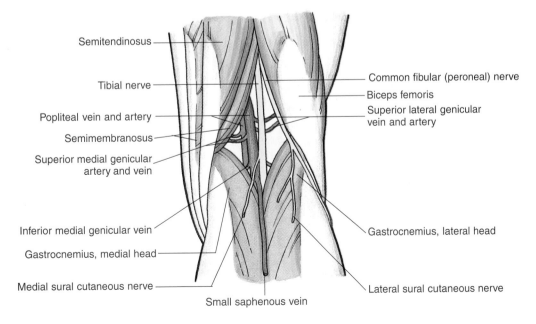

1. Relax the knee (flex it slightly) and incise the deep fascia to reveal the **gastrocnemius.** Separate its two heads inferiorly with your hands or a scalpel and expose the **soleus, popliteus,** and **plantaris muscles.** G5.42/C515-517/ R459-460/N481-482

2. Observe that the sciatic nerve divides into its **common peroneal** and **tibial nerve** divisions usually just above the knee.

3. Open the popliteal vascular fascial sheath and identify the **popliteal vein** and **artery.** From your atlas, appreciate the extensive anastomoses of arteries around the knee and identify the **superior lateral** and **medial genicular arteries.**

B. Anterior Leg and Dorsal Foot

1. On the skeleton, identify the following landmarks: A2. 45:46-56:52/ G5.1/ C589-591/ R416-419/ N478, 488
   a. **Medial condyle of tibia**
   b. **Lateral condyle of tibia**
   c. **Head of fibula**
   d. **Medial malleolus.** Large medial prominence at the ankle.
   e. **Lateral malleolus.** Lateral prominence at the ankle.
   f. Seven **tarsal bones: talus** (L. ankle), **calcaneus** (L. heel), **navicular** (L. little ship), **cuboid** (Gr. cube-shaped), and **three cuneiforms** (L. wedge-shaped).
   g. On the **calcaneus,** identify the **tuberosity** and **sustentaculum tali**
   h. Examine the five **metatarsals** and the **phalanges** (first toe only has two while other toes have three).

2. The anterior compartment of leg is the extensor (dorsiflexor) compartment (Table 5.5) (Fig. 5.7). Remove the deep fascia and note that some muscle fibers of the **tibialis anterior** arise from the deep fascia. At the ankle, observe thickenings of the deep fascia which keep the muscle tendons separate. These are the **superior** and **inferior extensor retinacula.** Just above the ankle, from medial to lateral, identify the **tibialis anterior, extensor hallucis** (great toe) **longus, deep peroneal (fibular) nerve** and **anterior tibial vessels,** and the **extensor digitorum longus muscle.** Pull on the muscle tendons to observe their actions. A2. 1:29:30-1:41:42/G5.62-5.63/C520-525/ R438, 466/N484-485

3. Follow the **anterior tibial artery** and **deep peroneal nerve** into the dorsum of the foot. Identify the **dorsalis pedis** (L., foot) **artery,** a continuation of the anterior tibial artery. Smaller

TABLE 5.5
MUSCLES IN ANTERIOR AND LATERAL COMPARTMENTS OF LEG

Muscle	Proximal Attachment	Distal Attachment	Innervation[a]	Main Actions
Anterior Compartment				
Tibialis anterior	Lateral condyle and superior half of lateral surface of tibia	Medial and inferior surfaces of medial cuneiform and base of first metatarsal	Deep fibular (peroneal) n. (**L4** and L5)	Dorsiflexes ankle and inverts foot
Extensor hallucis longus	Middle part of anterior surface of fibula and interosseous membrane	Dorsal aspect of base of distal phalanx of great toe (hallux)		Extends great toe and dorsiflexes ankle
Extensor digitorum longus	Lateral condyle of tibia and superior three-fourths of anterior surface of interosseous membrane and fibula	Middle and distal phalanges of lateral four digits	Deep fibular (peroneal) n. (L5 and S1)	Extends lateral four digits and dorsiflexes ankle
Fibularis (peroneus) tertius	Inferior third of anterior surface of fibula and interosseous membrane	Dorsum of base of fifth metatarsal		Dorsiflexes ankle and aids in eversion of foot
Lateral Compartment				
Fibularis (peroneus) longus	Head and superior two-thirds of lateral surface of fibula	Base of first metatarsal and medial cuneiform	Superficial fibular (peroneal) nerve (**L5, S1,** and **S2**)	Evert foot and weakly plantarflex ankle
Fibularis (peroneus) brevis	Inferior two-thirds of lateral surface of fibula	Dorsal surface of tuberosity on lateral side of base of fifth metatarsal		

[a]See Table 5.1 for explanation of segmental innervation.

branches of the anterior tibial artery on the foot include the arcuate artery, medial and lateral malleolar arteries, tarsal branches, and the dorsal digital arteries. The deep peroneal nerve supplies anterior compartment muscles and also supplies two short toe extensors, the **extensor digitorum brevis** and **extensor hallucis brevis**. A2. 1:43:10-1:47:56/ G5.64/ C525/ R466, 469/ N485, 506

C. Lateral Leg and Ankle

1. Identify the **peroneus longus** and **brevis muscles** (Fig. 5.8). The brevis inserts into the tuberosity of the 5th metatarsal

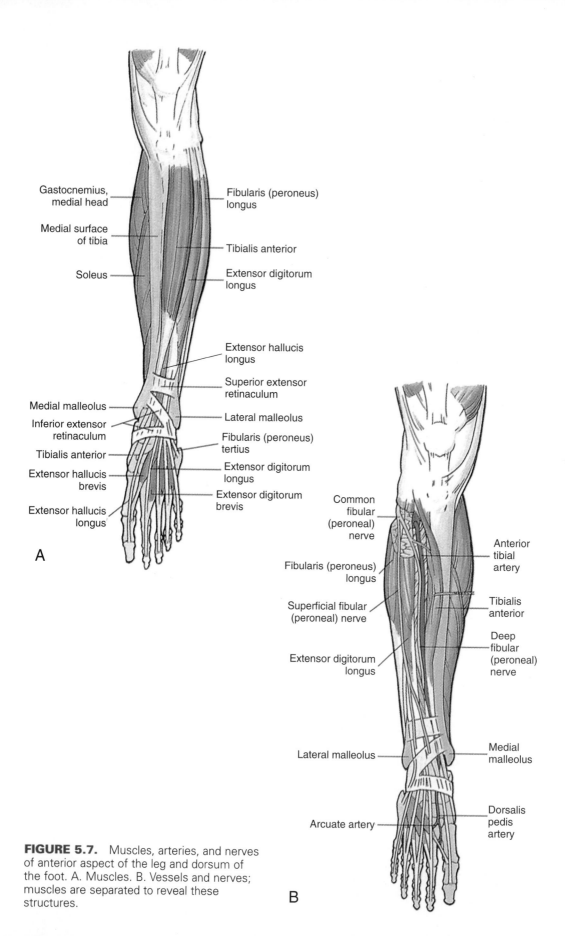

FIGURE 5.7. Muscles, arteries, and nerves of anterior aspect of the leg and dorsum of the foot. A. Muscles. B. Vessels and nerves; muscles are separated to reveal these structures.

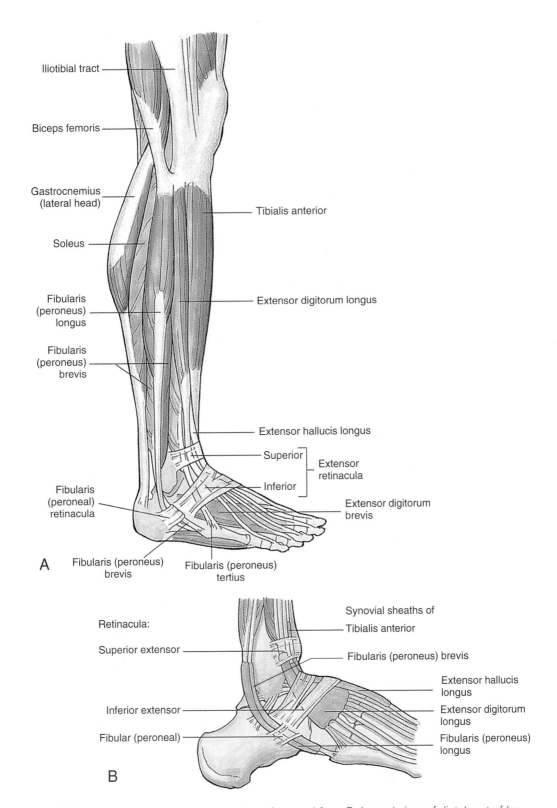

FIGURE 5.8. A. Lateral view of muscles of leg and foot. B. Lateral view of distal part of leg and proximal foot showing retinacula and synovial sheaths of tendons at the ankle.

bone while the longus tendon hooks around the cuboid and travels deep within the sole of the foot to insert into the 1st metatarsal bone (Table 5.5). The **superior** and **inferior peroneal retinacula** retain the tendons posterior to the lateral malleolus and calcaneus, respectively.   **G5.62-5.64/ C527/ R435, 466/N486**

2. Follow the **common peroneal nerve** and see its division into the **deep peroneal** and **superficial peroneal** branches. Note that the common peroneal nerve is vulnerable to injury where it wraps around the fibular head and neck superficially before piercing the peroneus longus muscle. The superficial branch of this nerve innervates the lateral compartment muscles. **G5.64/ C525/ R466/ N506**

D. Posterior Leg and Medial Ankle. The "calf" muscles are divided into three superficial muscles and four deep muscles. They are innervated by the tibial nerve and receive blood from the posterior tibial artery (Table 5.6).

1. Turn the cadaver to the prone position. Identify the two heads of the **gastrocnemius,** the **soleus,** and the **plantaris muscles.** The **calcaneal tendon** (Achilles tendon) is the fused tendons of the gastrocnemius and soleus.   **G5.70/ C540/ R433-434/ N481**

2. Cut the calcaneal tendon several centimeters superior to its insertion, reflect it superiorly, and pass two fingers superiorly between the soleus and intermuscular septum. With a scalpel, cut the soleus from the tibia but leave it attached to the fibula. Reflect the muscle laterally and view the **posterior tibial vessels** and **tibial nerve** (Fig. 5.9A).   **A2. 1:43:10-1:52:46/ G5.71/ C544/ R443, 462-463/ N482-483**

3. Identify the **flexor hallucis longus** (FHL) (lies laterally), the **flexor digitorum longus** (FDL) (lies medially), and the **tibialis posterior** (TP) (lies in the middle). A fourth muscle is the **popliteus,** which lies in the popliteal fossa under the gastrocnemius. Follow the tendons of the FHL, FDL, and TP around the medial side of the ankle (Fig. 5.9B). The tendon of the FHL runs in a groove below the sustentaculum tali, whereas the tendon of the FDL and TP cross each other in the sole of the foot. At the ankle, confirm the saying "Tom, Dick And Very Nervous Harry" (order of structures from anterior to posterior behind the medial malleolus: **TP** tendon, **FDL** tendon, **A**rtery, **V**ein, **N**erve, **FHL** tendon) (Fig. 5.9B). Study the actions of these muscles (Table 5.6).   **G5.71-5.74/ C544-548/ R436-437/ N483, 493**

4. Dissect and clean the **posterior tibial artery** and **tibial nerve** along their course. Identify the largest branch of the posterior

TABLE 5.6
MUSCLES IN POSTERIOR COMPARTMENT OF LEG

Muscle	Proximal Attachment	Distal Attachment	Innervation[a]	Main Actions
Superficial Muscles				
Gastrocnemius	*Lateral head:* lateral aspect of lateral condyle of femur *Medial head:* popliteal surface of femur, superior to medial condyle	Posterior surface of calcaneus via calcaneal tendon (tendo calcaneus)	Tibial nerve (S1 and **S2**)	Plantarflexes ankle, raises heel during walking, and flexes leg at knee joint
Soleus	Posterior aspect of head of fibula, superior fourth of posterior surface of fibula, soleal line, and medial border of tibia			Plantarflexes ankle and steadies leg on foot
Plantaris	Inferior end of lateral supracondylar line of femur and oblique popliteal ligament			Weakly assists gastrocnemius in plantarflexing ankle and flexing knee
Deep Muscles				
Popliteus	Lateral epicondyle of femur and lateral meniscus	Posterior surface of tibia, superior to soleal line	Tibial nerve (**L4, L5,** and S1)	Weakly flexes knee and unlocks it
Flexor hallucis longus	Inferior two-thirds of posterior surface of fibula and inferior part of interosseous membrane	Base of distal phalanx of great toe (hallux)	Tibial nerve (**S2** and S3)	Flexes great toe at all joints and plantarflexes ankle; supports medial longitudinal arch of foot
Flexor digitorum longus	Medial part of posterior surface of tibia inferior to soleal line, and from fascia covering tibialis posterior	Bases of distal phalanges of lateral four digits		Flexes lateral four digits and plantarflexes ankle; supports longitudinal arches of foot
Tibialis posterior	Interosseous membrane, posterior surface of tibia inferior to soleal line, and posterior surface of fibula	Tuberosity of navicular, cuneiform, and cuboid and bases of second, third, and fourth metatarsals	Tibial nerve (L4 and L5)	Plantarflexes ankle and inverts foot

[a]See Table 5.1 for explanation of segmental innervation.

tibial artery, the **peroneal artery** between the TP and FHL muscles, or within the FHL (Fig. 5.9A).

5. Review the leg in cross section and learn the muscles by their compartmental arrangement and actions. **G5.102/C614-617/ R471/ N487**

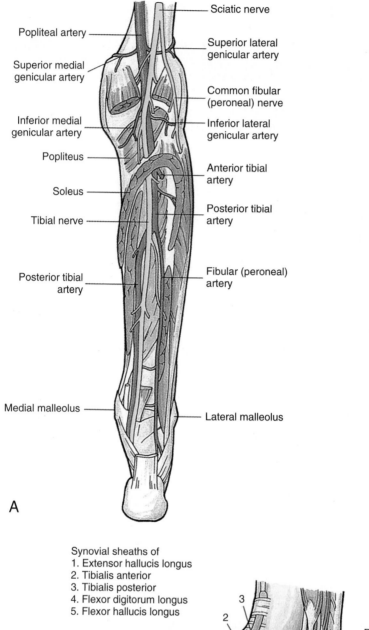

Sciatic nerve

Popliteal artery

Superior lateral
genicular artery

Superior medial
genicular artery

Common fibular
(peroneal) nerve

Inferior medial
genicular artery

Inferior lateral
genicular artery

Popliteus

Anterior tibial
artery

Soleus

Posterior tibial
artery

Tibial nerve

Posterior tibial
artery

Fibular (peroneal)
artery

Medial malleolus

Lateral malleolus

A

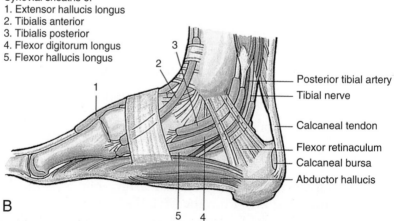

Synovial sheaths of
1. Extensor hallucis longus
2. Tibialis anterior
3. Tibialis posterior
4. Flexor digitorum longus
5. Flexor hallucis longus

Posterior tibial artery

Tibial nerve

Calcaneal tendon

Flexor retinaculum

Calcaneal bursa

Abductor hallucis

B

FIGURE 5.9.   A. Deep dissection of posterior aspect of leg showing nerves and arteries. Soleus is largely cut away. B. Medial view of distal part of leg, ankle, and foot showing retinacula and synovial sheaths of tendons at the ankle.

IV. SOLE OF THE FOOT

Learning Objectives

- Describe the actions of the intrinsic foot muscles.
- List the vascular supply and innervation to these muscles.
- Identify the important surface features of the ankle and dorsum of the foot in your textbook, atlas, and on yourself.

Key Concepts

- Medial and lateral plantar neurovascular bundles
- Plantar and dorsal interosseous muscle actions
- Parallels between upper and lower extremity

A. Introduction. Some courses will not require this dissection. Please check with your instructor.

From your textbook, note that the foot is arched longitudinally and possesses weight-bearing points. A tough plantar aponeurosis (fascia) protects the sole and helps to maintain the arch. Beneath the aponeurosis lie *four* layers of muscles. Their blood supply comes from the posterior tibial artery and their innervation from the terminal branches of the tibial nerve (medial and lateral plantar nerves).

B. First Layer

1. Remove the skin over the sole of the foot and cut the **plantar aponeurosis** longitudinally from the base of the toes to the heel (Fig. 5.10A). Carefully reflect the aponeurosis to reveal the underlying structures. Identify the **flexor digitorum brevis, abductor hallucis,** and **abductor digiti minimi** (Fig. 5.10B). Cut the flexor digitorum brevis close to the calcaneus and reflect it anteriorly toward the toes. Demonstrate the **medial** and **lateral plantar arteries** and **nerves** in the sole and observe their continuity with the posterior tibial artery and tibial nerve in the leg (Fig. 5.10C) (Table 5.7). **A2. 2:02:18-2:16:38/ G5.75-5.77/ C549-551/ R439/N496-497**

C. Second Layer

1. Identify the **quadratus plantae** and note its insertion into the tendon of the FDL. Identify four delicate **lumbrical muscles** which arise from the tendons of the FDL (Fig. 5.10C). **G5.78/ C552/ R440/ N498**

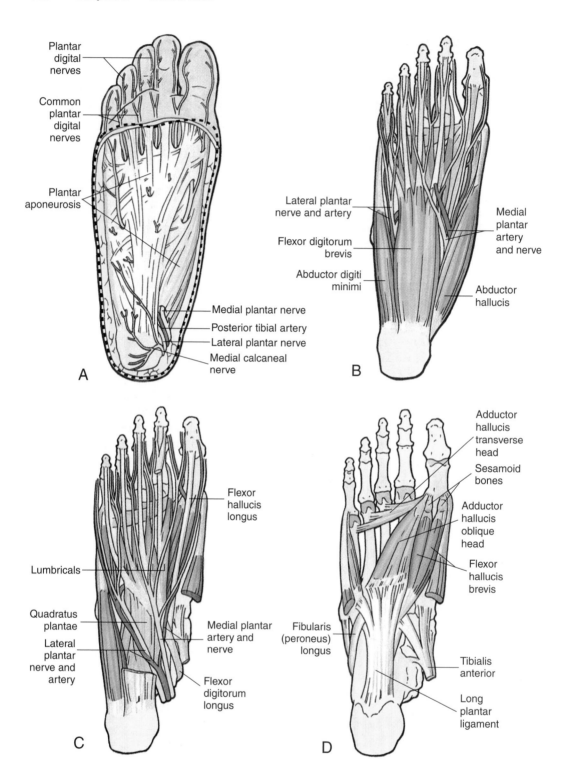

FIGURE 5.10. Dissections of sole of foot. A. Superficial dissection showing plantar aponeurosis. B. First layer of plantar muscles, digital nerves, and arteries. C. Second layer of plantar muscles. D. Third layer of plantar muscles.

TABLE 5.7
MUSCLES IN SOLE OF FOOT

Muscle	Proximal Attachment	Distal Attachment	Innervation[a]	Main Actions
First Layer				
Abductor hallucis	Medial tubercle of tuberosity of calcaneus, flexor retinaculum, and plantar aponeurosis	Medial side of base of proximal phalanx of first digit	Medial plantar nerve (S2 and **S3**)	Abducts and flexes first digit (great toe, hallux)
Flexor digitorum brevis	Medial tubercle of tuberosity of calcaneus, plantar aponeurosis, and intermuscular septa	Both sides of middle phalanges of lateral four digits		Flexes lateral four digits
Abductor digit minimi	Medial and lateral tubercles of tuberosity of calcaneus, plantar aponeurosis, and intermuscular septa	Lateral side of base of proximal phalanx of fifth digit	Lateral plantar nerve (S2 and **S3**)	Abducts and flexes fifth digit
Second Layer				
Quadratus plantae	Medial surface and lateral margin of plantar surface of calcaneus	Posterolateral margin of tendon of flexor digitorum longus	Lateral plantar nerve (S2 and **S3**)	Assists flexor digitorum longus in flexing lateral four digits
Lumbricals	Tendons of flexor digitorum longus	Medial aspect of expansion over lateral four digits	*Medial one:* medial plantar nerve (S2 and **S3**) *Lateral three:* lateral plantar nerve (S2 and **S3**)	Flex proximal phalanges and extend middle and distal phalanges of lateral four digits
Third Layer				
Flexor hallucis brevis	Plantar surfaces of cuboid and lateral cuneiforms	Both sides of base of proximal phalanx of first digit	Medial plantar nerve (S2 and **S3**)	Flexes proximal phalanx of first digit
Adductor hallucis	*Oblique head:* bases of metatarsals 2–4 *Transverse head:* plantar ligaments of metatarsophalangeal joints	Tendons of both heads attach to lateral side of base of proximal phalanx of first digit	Deep branch of lateral plantar nerve (S2 and **S3**)	Adducts first digit; assists in maintaining transverse arch of foot
Flexor digit minimi brevis	Base of fifth metatarsal	Base of proximal phalanx of fifth digit	Superficial branch of lateral plantar nerve (S2 and **S3**)	Flexes proximal phalanx of fifth digit, thereby assisting with its flexion

TABLE 5.7 (Continued)
MUSCLES IN SOLE OF FOOT

Muscle	Proximal Attachment	Distal Attachment	Innervation[a]	Main Actions
Fourth Layer				
Plantar interossei (three muscles)	Bases and medial sides of metatarsals 3–5	Medial sides of bases of proximal phalanges of third to fifth digits	Lateral plantar nerve (S2 and **S3**)	Adduct digits (2–4) and flex metatarsophalangeal joints
Dorsal interossei (four muscles)	Adjacent sides of metatarsals 1–5	*First:* medial side of proximal phalanx of second digit *Second to fourth:* lateral sides of second to fourth digits		Abduct digits (2–4) and flex metatarsophalangeal joints

D. Third Layer

1. Cut between the FDL tendon and the quadratus plantae and reflect the FDL tendons and lumbricals forward. Identify the two heads of the **flexor hallucis brevis,** and a **sesamoid bone** in each of its tendons (not always present) (Fig. 5.10D). The tendon of the FHL runs between the two sesamoid bones. Identify the **adductor hallucis** (transverse and oblique heads) and find the **flexor digiti minimi.**   G5.80/ C554-556/ R441/ N498-499

E. Fourth Layer

1. From your atlas, appreciate that this layer consists of the **dorsal** (four) and **plantar** (three) **interossei muscles,** which lie between adjacent metatarsals. Dorsal interossei abduct (DAB) and plantar interossei adduct (PAD) the toes. The second toe is the reference axis for defining abduction and adduction of the toes (in the hand, the middle finger is the reference axis). The arrangement and action of the interossei is similar to that in the hand.

Identify the **dorsal interossei** on the dorsum of the foot and find at least one **plantar interosseus muscle** deep in the sole. Also deep within the sole, find the **peroneus longus tendon** passing to the base of the 1st metatarsal bone. Look for the **plantar arterial arch** and appreciate that it has anastomotic connections via perforating branches with the dorsalis pedis artery.   A2. 2:17:53-2:25:50/G5.80-5.81/C553-556/ R472-474/ N500-501

2. If you have already dissected the upper limb, think about parallels between the upper and lower extremity with respect to the muscle compartments, patterns of innervation, blood supply, and skeletal framework. Differences also exist related to the different functions of the two limbs. If you have not dissected the upper limb yet, when you do, recall the similarities between the lower limb and the organization of the upper limb.

V. JOINTS OF THE LOWER LIMB

Learning Objectives

- Identify the key ligaments of the joints and understand the important role that muscles crossing the joint may play in stability and movements of the joint.
- List the vascular supply to the hip, knee, and ankle joints.
- From your textbook, learn about common knee injuries and describe how the anterior and posterior cruciate ligaments stabilize the knee.
- List the ligaments that are most vulnerable in ankle inversion injuries.

Key Concepts

- Types of joints
- Nerves supplying muscles that act on a joint, also innervate the joint
- Joint ligaments and their relationship to joint stability or mobility

---

A. Hip Joint

1. *Perform this dissection on one hip only, saving the other side for review.* Remove all of the muscles on the anterior aspect of the hip joint (sartorius, rectus femoris, and pectineus), reflect the femoral vessels, and sever the tendon of the iliopsoas muscle close to the lesser trochanter.

2. Identify the strong inverted Y-shaped **iliofemoral ligament** anteriorly. Note that hip extension stretches this ligament (prevents over extension during standing). Open the **capsule** of the joint anteriorly and observe the articular surface of the head of the femur (Fig. 5.11).   **A2. 00:49-11:08/ G5.33-5.34/ C563-568/ R420-421/N454**

3. Turn the cadaver into the prone position, remove the piriformis, obturator internus, the gemelli muscles, quadratus femoris, obturator externus, gluteus medius and minimus

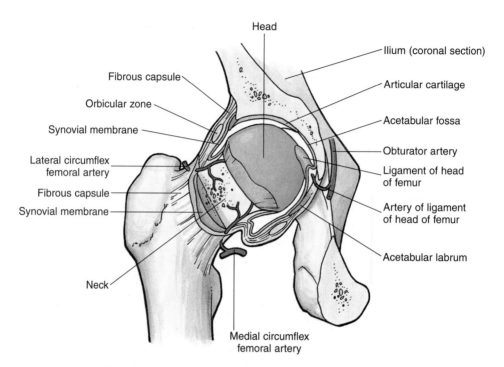

FIGURE 5.11. Blood supply of femoral head and neck.

muscles. Clean the posterior capsule and identify the **is-chiofemoral ligament** (also prevents hyperextension).

4. Open the capsule and cut the **ligament of the head of the femur (ligamentum teres)**. Identify the smooth lunate articular surface. Study the blood supply to the hip joint from the circumflex femoral branches and the **artery to the femoral head,** which arises from the obturator artery (Fig. 5.11).

B. Knee Joint

1. *Perform this dissection on only one limb.* On the medial side, detach the tendons of the sartorius, gracilis, and semitendinosus (**pes anserinus**) from their insertions. Deep to the region where these three muscles cross the joint, identify the superficial portion of the **tibial (medial) collateral ligament.** The deeper portion of this ligament is attached to the medial meniscus (very important) (Fig. 5.12 A and B). **A2. 45:46-58:20/ G5.46/ C570/ R422-424/ N472, 475**

2. On the lateral side, cut the biceps tendon close to its insertion on the head of the fibula. Identify the **fibular (lateral) collateral ligament.** Observe that the **popliteus tendon** actually passes in the space between the ligament and the meniscus. **G5.47/ C575-576/ R425/N472, 475**

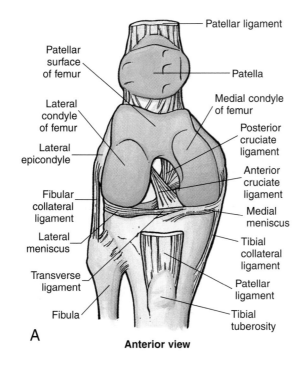

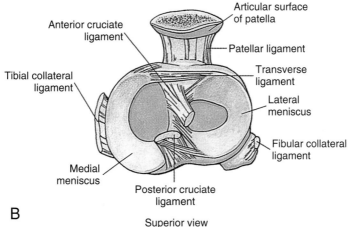

FIGURE 5.12.   A. Anterior view of the knee joint. B. Superior view of the knee joint showing ligaments and menisci.

3. Anteriorly, cut into the quadriceps tendon just inferior to the patella to reveal the underlying synovial capsule but do not completely cut across the quadriceps tendon.

Identify the **suprapatellar** (quadriceps) **bursa.** Now, cut the patellar tendon just above its attachment to the tibial tuberosity and reflect it superiorly (Fig. 5.12A).   **G5.49, 5.56/ C572/ R422-423/ N473**

4. Posteriorly, remove the vessels from the popliteal fossa, cut hamstring muscles superior to the knee, and free the plantaris and both heads of the gastrocnemius from the joint capsule. Remove the popliteus muscle and open the joint capsule. Identify the **posterior cruciate ligament.** Then open the synovial cavity extending between the femoral condyles, posteriorly and anteriorly, and note that the cruciate ligaments lie outside of the synovial cavity (Fig. 5.12).   **G5.53-5.57/ C576-577/ R422/ N474-476**

5. Developmentally, the knee joint had medial and lateral halves separated by membranes of the intercondylar septum. Anteriorly, the septum is incomplete and forms the infrapatellar synovial fold. Posteriorly, the septum is complete. Cut this fold and study the **anterior** and **posterior cruciate ligaments** by flexing and extending the knee to determine the function of these ligaments. Cut just the anterior cruciate ligament and see what happens during flexion of the knee. Study the C-shaped **medial meniscus** and the O-shaped **lateral meniscus,** which is more mobile (no attachment to the lateral collat-

FIGURE 5.13.   Ligaments of ankle joint. Top view shows the lateral side and the bottom view the medial side.

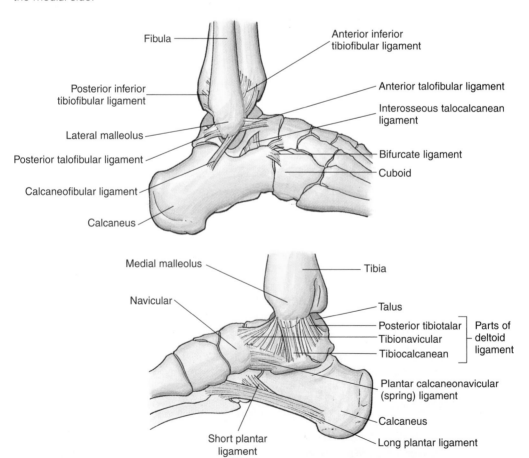

eral ligament) (Fig. 5.12B). Read about knee injuries in your textbook.   **G5.54/ C585-588/ R422/ N474**

C. **Ankle Joint**

1. To display the medial aspect of the ankle joint, cut the flexor digitorum longus tendon and displace, but do not cut the tibialis posterior tendon anteriorly. Clean and define the **medial** or **deltoid ligament** which is triangular in shape (Fig. 5.13). **A2. 1:14:53-1:29:30/ G5.88/C606/R425-426/ N491**

2. Laterally, pull the peroneus longus and brevis tendons anteriorly and clean the **calcaneofibular** and **anterior talofibular ligaments.** Make sure that the superior and inferior peroneal retinacula are slit open. Also identify the strong **anterior inferior tibiofibular ligament,** which helps to hold the tibia and fibula together.   **G5.87-5.89/ C604/ R425-426/N490-492**

3. On the sole of the foot, identify the "**spring ligament**" or **plantar calcaneonavicular ligament,** which, along with the tibialis posterior tendon, are necessary for the support of the head of the talus (Fig. 5.13). Also, identify the **long** and **short plantar ligaments** (Figs. 5.10D and 5.13).   **A2. 1:51:46-2:02:18/ G5.93/ C610-611/ R426/ N492**

6

UPPER LIMB

I. SHOULDER

Learning Objectives

- Throughout Chapter 6, identify structures in bold print unless instructed to do otherwise.
- Know which veins contribute to the superficial venous return of the upper limb.
- Identify the scapular features and the muscles that attach to it.
- Describe the "rotator cuff" and state why it is important for shoulder stability.
- List the muscles innervated by the axillary nerve and suprascapular nerve, and describe the functional deficits that might occur if these nerves were damaged.

Key Concepts

- Limb rotation during development
- Superficial and deep venous return
- Muscle compartments
- Rotator cuff

A. Introduction

If the Thorax (Chapter 1) has been dissected previously, then the pectoral muscles have already been identified. If the Back (Chapter 4) has been dissected previously, then the superficial group of back muscles that connect the upper limb to the vertebral column have been identified. If you are assigned the dissection of the

121

Upper Limb before the Thorax or Back, then the muscles mentioned above must be studied initially.

1. First, examine the entire upper limb noting especially the superficial venous drainage (Fig. 6.1). Focus on points where you might introduce a needle to draw blood or start intravenous fluids. Also note where the arteries are superficial and you can feel a pulse.   **G6.4-6.6/ C37, 43/ R376/ N442, 448-449**

2. Referring to Figure 6.2, make the designated skin incisions and remove the skin and superficial fascia (tela). First make incision A-B. Then complete the circular incisions C-C, D-D, E-E, and F-F (Fig. 6.2) (Some instructors may prefer that the

FIGURE 6.1.   Superficial veins and lymphatic drainage.

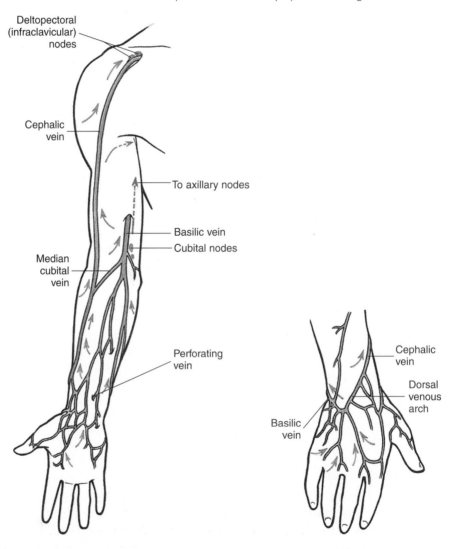

Anterior view   **Dorsal hand**

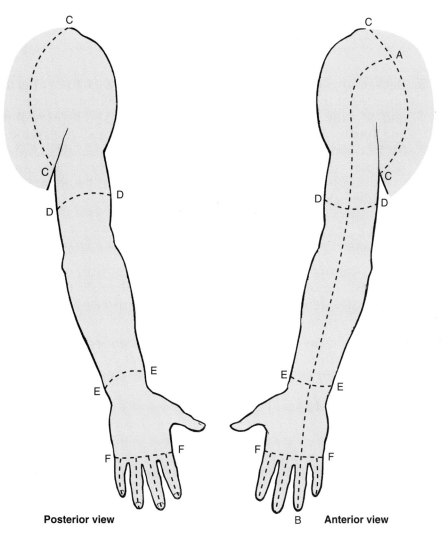

Posterior view   B   Anterior view

FIGURE 6.2.   Upper limb. Dashed lines show skin incisions.

skin be removed in vertical strips from proximal to distal "with the grain" in terms of the vertical orientation of the major veins and cutaneous nerves. Please check with your instructor). In so doing, superficial veins and cutaneous nerves will be removed, so study these in your atlas. If the hand is clenched or the fingers curled, have your lab partner straighten them out while you make the incisions in the palm and along the fingers. On the fingers, be careful not to cut too deep. As you remove the skin, look for the **basilic vein, cephalic vein** and **median cubital vein.** Once the skin and tela is removed, you will encounter the thick investing fascia of the arm, the **brachial fascia,** and forearm, the **antebrachial fascia.** At the wrist, note the extensor and flexor retinaculum, and in the cubital region, the bicipital aponeurosis. On the dorsum of the hand, note the **superficial dorsal veins.** **G6.4/ C44, 76/ R376/ N448-449**

B. **Back and Shoulder Regions.** Understand that the superficial muscles of the back are muscles associated with the upper limb and include the **trapezius, latissimus dorsi,** the **rhomboids** and **levator scapulae.** If you have not previously dissected the back (Chapter 4), these muscles will need to be dissected first (Chapter 4, Unit 1-Back, Section B). Review the actions and innervation of these muscles. **A1. 9:51-20:36/ G6.25/C629/ R360-361/N160, 163, 395**

1. **Bony Landmarks.** Referring to the skeleton identify the following: **A1. 1:27-8:13/ G6.1/ C92-94, 104-107/ R349-351/ N392-393**

   a. **Scapula** and its **acromion, spine, supraspinous fossa, infraspinous fossa, glenoid cavity, supraglenoid tubercle, infraglenoid tubercle, coracoid process,** and **scapular notch.**

   b. **Humerus** and its **head, greater tubercle, lesser tubercle, intertubercular sulcus** (bicipital groove), **deltoid tuberosity,** and **sulcus for the radial nerve (spiral groove).**

2. **Deltoid Muscle.** Place the cadaver in the prone position. Dissection will be easier if the arm is abducted slightly and the shoulder is allowed to fall forward. Define the borders of the **deltoid muscle** and detach the muscle from the spine and the acromion of the scapula, leaving the muscle attached to the clavicle. Reflecting the deltoid anteriorly, identify the **axillary nerve** and **posterior circumflex humeral artery** on its deep surface. Note that these structures pass through the **quadrangular space.** **A1. 17:58-20:36/ G6.29-6.30/ C31, 52/ R379/N395, 397, 400**

3. **Triceps Muscle.** Identify the **long head of the triceps brachii** and note that it passes between the **teres minor** and **teres major muscles** (Table 6.1). Using your fingers, separate the long head from the **lateral head.** Define the triangular interval between the two heads and identify the **radial nerve** and **deep brachial artery** (and veins) in the space. The nerve and vessels lie in the **spiral groove of the humerus.** Note that veins accompany the deep arteries throughout the upper limb, but our focus will be on dissecting and identifying the key arteries. **G6.31/ C55, 57/ R379-381/ N403, 405**

4. **Supraspinatus and Infraspinatus Muscles.** Find these two muscles in their respective fossae (Fig. 6.3). Reflect the trapezius anteriorly to expose these muscles. Remove the investing (deep) fascia over these muscles and cut vertically across the supraspinatus muscle about 5 cm lateral to the superior angle of the scapula (medial to suprascapular

TABLE 6.1
SUPERFICIAL BACK, SCAPULAR, AND ARM MUSCLES

Muscle	Proximal Attachment	Distal Attachment	Innervation	Main Actions
Trapezius	Medial third of superior nuchal line; external occipital protuberance, ligamentum nuchae, and spinous processes of C7–T12 vertebrae	Lateral third of clavicle, acromion, and spine of scapula	Spinal root of accessory n. (CN XI) and cervical nn. (C3 and C4)	Elevates, retracts, and rotates scapula; superior fibers elevate, middle fibers retract, and inferior fibers depress scapula; superior and inferior fibers act together in superior rotation of scapula
Latissimus dorsi	Spinous processes of inferior six thoracic vertebrae, thoracolumbar fascia, iliac crest, and inferior three or four ribs	Floor of intertubercular groove of humerus	Thoracodorsal n. (**C6, C7,** and C8)	Extends, adducts, and medially rotates humerus; raises body toward arms during climbing
Levator scapulae	Posterior tubercles of transverse processes of C1–C4 vertebrae	Superior part of medial border of scapula	Dorsal scapular (C5) and cervical (C3 and C4) nn.	Elevates scapula and tilts its glenoid cavity inferiorly by rotating scapula
Rhomboid minor and major	*Minor:* ligamentum nuchae and spinous processes of C7 and T1 vertebrae *Major:* spinous processes of T2–T5 vertebrae	Medial border of scapula from level of spine to inferior angle	Dorsal scapular n. (C4 and **C5**)	Retracts scapula and rotates it to depress glenoid cavity; fixes scapula to thoracic wall
Deltoid	Lateral third of clavicle, acromion, and spine of scapula	Deltoid tuberosity of humerus	Axillary n. (**C5** and C6)	*Anterior part:* flexes and medially rotates shoulder *Middle part:* abducts shoulder *Posterior part:* extends and laterally rotates shoulder
Supraspinatus[a]	Supraspinous fossa of scapula	Superior facet on greater tubercle of humerus	Suprascapular n. (C4, **C5,** and C6)	Helps deltoid to abduct shoulder and acts with rotator cuff muscles[a]
Infraspinatus[a]	Infraspinous fossa of scapula	Middle facet on greater tubercle of humerus	Suprascapular n. (**C5** and C6)	Laterally rotate shoulder; help to hold humeral head in glenoid cavity of scapula
Teres minor[a]	Superior part of lateral border of scapula	Inferior facet on greater tubercle of humerus	Axillary n. (**C5** and C6)	

[a]Collectively, the supraspinatus, infraspinatus, teres minor, and subscapularis muscles are referred to as the rotator cuff muscles. Their prime function during all movements of shoulder joint is to hold the head of humerus in the glenoid cavity of the scapula.

Continued

TABLE 6.1 (Continued)
SUPERFICIAL BACK, SCAPULAR, AND ARM MUSCLES

Muscle	Proximal Attachment	Distal Attachment	Innervation	Main Actions
Teres major	Dorsal surface of inferior angle of scapula	Medial lip of intertubercular groove of humerus	Lower subscapular n. (**C6** and C7)	Adducts and medially rotates shoulder
Subscapularis[a]	Subscapular fossa	Lesser tubercle of humerus	Upper and lower subscapular nn. (C5, **C6**, and C7)	Medially rotates shoulder and adducts it; helps to hold humeral head in glenoid cavity
Biceps brachii	*Short head:* tip of coracoid process of scapula *Long head:* supraglenoid tubercle of scapula	Tuberosity of radius and fascia of forearm via bicipital aponeurosis	Musculocutaneous n. (C5 and **C6**)	Supinates forearm and when it is supine, flexes elbow
Brachialis	Distal half of anterior surface of humerus	Coronoid process and tuberosity of ulna		Flexes elbow in all positions
Coracobrachialis	Tip of coracoid process of scapula	Middle third of medial surface of humerus	Musculocutaneous n. (C5, **C6,** and C7)	Helps to flex and adduct shoulder
Triceps brachii	*Long head:* infraglenoid tubercle of scapula *Lateral head:* posterior surface of humerus, superior to radial groove *Medial head:* posterior surface of humerus, inferior to radial groove	Proximal end of olecranon of ulna and fascia of forearm	Radial n. (C6, **C7,** and **C8**)	Extends elbow; it is chief extensor of elbow; long head steadies head of abducted humerus
Anconeus	Lateral epicondyle of humerus	Lateral surface of olecranon and superior part of posterior surface of ulna	Radial n. (C7, C8, and T1)	Assists triceps in extending elbow; stabilizes elbow joint; abducts ulna during pronation

[a]Collectively, the supraspinatus, infraspinatus, teres minor, and subscapularis muscles are referred to as the rotator cuff muscles. Their prime function during all movements of shoulder joint is to hold the head of humerus in the glenoid cavity of the scapula.

notch), reflect it laterally, and clean the **suprascapular nerve** and **artery** (nerve passes under the **suprascapular ligament**). Also cut vertically across the infraspinatus muscle 5 cm lateral to the vertebral border, peel the muscle laterally, and follow the suprascapular artery and nerve into the infraspinous fossa. Note from your atlas and textbook that the scapular region has an extensive collateral circulation,

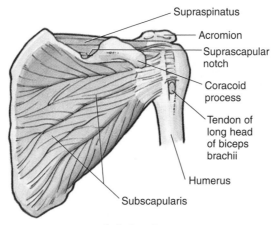

Anterior view

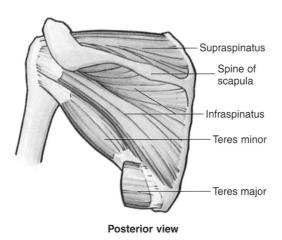

Posterior view

FIGURE 6.3. Rotator cuff muscles.

which is important surgically.   **A1. 10:24-13:39/ G6.30/ C31-34/ R381-383/N395-498**

5. **Rotator Cuff.** Note the insertions of the **supraspinatus, infraspinatus** and **teres minor muscles,** which fuse with the capsule of the shoulder joint (Fig. 6.3). These three muscle tendons and the insertion of the **subscapularis** form the rotator cuff. Study their importance in your textbook.   **G6.28/ C33/ R361/ N396**

6. If you have already dissected the lower limb, look for parallels in the skeletal framework, vascular supply, and muscle compartments between the lower and upper limbs. Although differences exist because of the different functions of the limbs, many similarities are present, and these similarities will help you organize your study of the limbs.

II. AXILLA AND ARM

Learning Objectives

- Identify the key anatomical features of the humerus.
- List the structures found in the axillary sheath.
- Describe the arterial anastomosis around the shoulder and state why it is important clinically.
- List the terminal branches of the brachial plexus and identify the spinal cord segments that contribute to their formation.
- Define the actions of the muscles of the arm on the shoulder and elbow joints.
- Label on a cross-section of the arm the functional compartments and vessels and nerves of each compartment.

Key Concepts

- Collateral circulation around joints
- Muscle compartments
- Axillary lymph nodes

A. Axilla. If you have not dissected the thorax (Chapter 1) previously, you will need to dissect the pectoral muscles at this time (Chapter 1, Unit 1-Thoracic Wall, Sections C-D). The axilla is the region between the arm and the chest (arm pit). Note its boundaries from your textbook and palpate them on yourself. **G6.15/ C19-21/ R386/ N399**

1. **Anterior Wall.** Again, note the **pectoralis major** and **minor muscles** and be sure they are reflected toward the arm.

2. **Posterior Wall.** Identify the **latissimus dorsi, teres major** and **subscapularis.**

3. **Medial Wall.** Identify the **serratus anterior muscle.** Note that this muscle covers the ribs and intercostal muscles. With your fingers, follow the muscle dorsally toward the medial margin of the scapula.

4. **Lateral Wall.** Identify the **intertubercular sulcus of the humerus** (bicipital groove).

5. **Axillary Contents**
   a. Identify the **long** and **short heads of the biceps brachii** (short head tendon lies medial to long head). Medially, observe the **coracobrachialis.** **G6.13/ C19-30/ R387/ N402, 404**

b. Identify the **axillary sheath,** a continuation of prevertebral fascia from the neck, which envelops the **axillary artery,** veins and brachial plexus of (ventral rami of C5-T1). Carefully open the axillary sheath, separate the brachial plexus from the axillary artery without damaging the nerves, and clean the artery (Fig. 6.4). The axillary artery begins at the first rib and ends as the brachial artery at the inferior border of the teres major muscle. Classically, the artery is divided into three parts: the first part extends from the 1st rib to the pectoralis minor, the second part lies deep to the pectoralis minor, and the third part extends from the pectoralis minor to the inferior border of the teres major.   **A1. 28:00-30:22/ G6.12-6.14/ C23/ R386-387/ N400**

c. From the first part of the axillary artery, find the **superior (supreme) thoracic artery** if possible (it is very small). From the second part, find the **thoracoacromial artery** and **lateral thoracic artery** (both arise deep to the pectoralis minor muscle) (Fig. 6.4). From the third part of the axillary artery, find the **subscapular artery** and its two terminal branches the **thoracodorsal artery** (to latissimus dorsi) and **circumflex scapular artery.** Next, find the larger **posterior circumflex humeral artery** (passes through quadrangular space) and the smaller **anterior circumflex humeral artery.** Note the important anastomoses that these arteries provide around the scapula. Also realize that numerous lymph nodes reside in the axilla and drain the upper limb and portions of the thoracic wall, including the breast. Look for nodes in the fatty substance of the axilla (see Fig. 1.2, Chapter 1).   **G6.5, 6.14/ C23/ R375, 382/ N398, 400, 405**

6. Axillary Contents/Brachial Plexus. Now completely open the axillary sheath and separate the cords of the brachial plexus. The axillary artery is surrounded by the three **cords of the plexus (lateral, medial** and **posterior)** (Fig. 6.5). Of the five terminal branches of the plexus, the **musculocutaneous nerve** is most lateral and enters the coracobrachialis muscle. Also from the lateral cord, arises part of the **median nerve,** which receives a contribution from the medial cord. The terminal branch of the medial cord is the **ulnar nerve.** The posterior cord gives rise to the **axillary nerve** (passes through quadrangular space) and the **radial nerve** (follows deep brachial artery around the humeral shaft posteriorly).   **A1. 30:23-35:55/ G6.20/ C26-28/ R389-390/ N400-401**

The medial cord also gives rise to a large **medial cutaneous nerve of the forearm.** The posterior cord gives rise to the **subscapular nerves,** which supply the subscapularis and teres major muscles. Also look for the nerve to the latissimus dorsi (**thoracodorsal nerve**) and nerve to the serratus anterior (**long thoracic nerve**) which arises from the first three roots of the

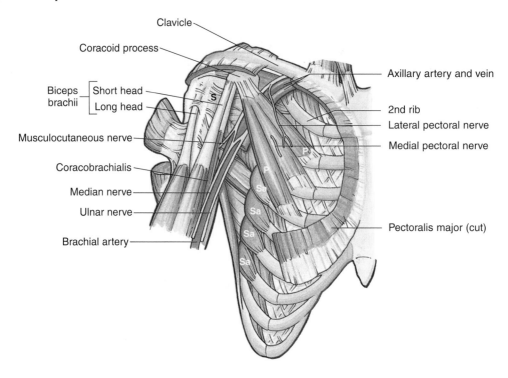

Clavicle

Coracoid process

Axillary artery and vein

Biceps brachii ⌈ Short head
              ⌊ Long head

2nd rib

Lateral pectoral nerve

Musculocutaneous nerve

Medial pectoral nerve

Coracobrachialis

Median nerve

Ulnar nerve

Pectoralis major (cut)

Brachial artery

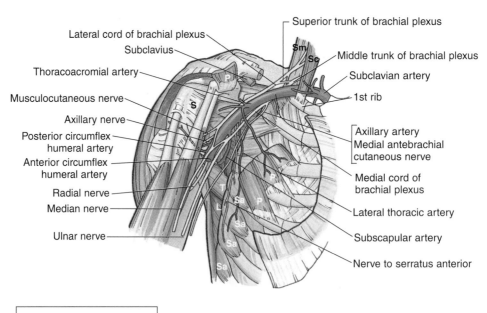

Lateral cord of brachial plexus

Superior trunk of brachial plexus

Subclavius

Thoracoacromial artery

Middle trunk of brachial plexus

Subclavian artery

Musculocutaneous nerve

1st rib

Axillary nerve

Posterior circumflex humeral artery

Axillary artery
Medial antebrachial cutaneous nerve

Anterior circumflex humeral artery

Radial nerve

Medial cord of brachial plexus

Median nerve

Lateral thoracic artery

Ulnar nerve

Subscapular artery

Nerve to serratus anterior

Key	
P	Pectoralis minor
S	Subscapularis
Sa	Serratus anterior
Sc	Scalenus anterior
Sm	Scalenus medius
T	Teres major
L	Latissimus dorsi

FIGURE 6.4. Axillary region. In the lower figure, both pectoral muscles have been cut and reflected.

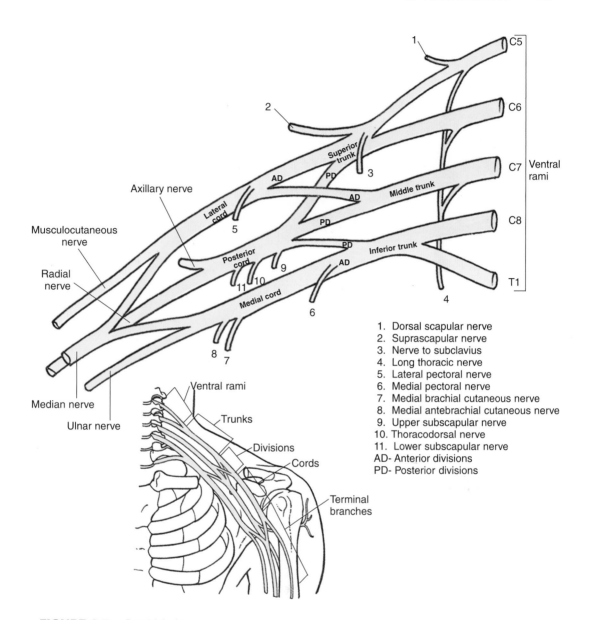

FIGURE 6.5. Brachial plexus.

plexus (C5-7). Now clean the **serratus anterior muscle** and **subscapularis muscle.**

B. Arm. The muscles of the arm are contained in two fascial compartments. The posterior compartment contains extensors, and the anterior compartment contains flexors of the elbow (Table 6.1).

1. Posterior Compartment. Identify the three heads of the **triceps,** the **long, lateral,** and **medial heads** (Fig. 6.6). Expose the

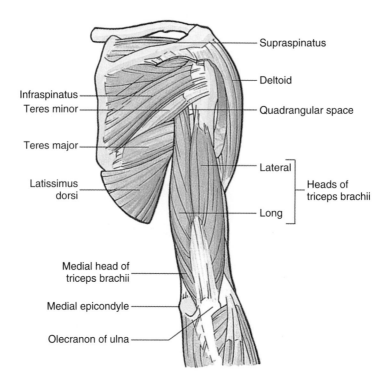

FIGURE 6.6. Posterior muscles of the shoulder and arm.

radial nerve and **deep brachial artery** by carefully cutting the lateral head of the triceps. Look for this neurovascular bundle in the spiral groove of the humerus, deep to the lateral head of the triceps.   A1. 55:15-59:07/G6.31/C54-57/ R379-380/ N403, 406

2. **Anterior Compartment.** Identify the **coracobrachialis, brachialis,** and **biceps brachii muscles** (Fig. 6.7). Trace the pathway of the **musculocutaneous nerve** through the coracobrachialis and between the biceps and brachialis muscles to terminate as the **lateral cutaneous nerve of the forearm**. Sever the biceps about 5 cm proximal to the cubital region, and reflect the muscle to better see the musculocutaneous nerve.   G6.21/ C48-51/ R364-365, 391/ N402, 406, 408

Trace and clean the **median** and **ulnar nerves** to the elbow.

Follow and clean the **brachial artery** from the lower border of the teres major muscle to its bifurcation in the cubital fossa into the **ulnar** and **radial arteries.** The brachial artery gives rise to the **deep brachial (profunda brachii)** and numerous muscular and collateral branches (ulnar collateral arteries), which we will not dissect. However, know why they may become important if vascular occlusion of the brachial artery occurs.   G6.61/ C57, 63/ R375/ N405, 417-418

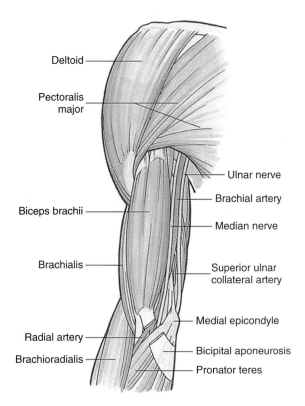

FIGURE 6.7. Anterior muscles of the shoulder and arm showing the brachial artery and associated nerves. The biceps tendon is cut.

III. FLEXOR REGION OF THE FOREARM

Learning Objectives

- Identify key features of the radius and ulna.
- Name the muscles of the flexor compartment of the forearm and describe their actions and innervation.
- At the wrist, identify the muscle tendons and median nerve as they pass through the "carpal tunnel."

Key Concepts

- Muscle compartments
- Carpal tunnel
- Pronation

A. Introduction. The muscles of the anterior forearm are divided into a superficial and deep group. The ulnar nerve and artery, and median nerve lie in a connective tissue septum that separates these two groups. Superficial flexors arise from the medial epicondyle of the humerus, whereas deep muscles arise from the radius and ulna.

B. Bony Landmarks

1. On a skeleton identify the following landmarks: A1. 41:11-52:45/ G6.1/ C104-111/ R351-353/ N407-409

a. **Humerus** and its **medial** and **lateral epicondyles, capitulum, trochlea,** and **olecranon fossa.**

b. **Radius** and its **head, neck, tuberosity, styloid process,** and **interosseous border.**

c. **Ulna** and its **head, olecranon,** and **interosseous border.**

C. Forearm Structures

1. Superficial Muscles and Vessels. Remove the investing (deep) fascia from the forearm and identify the superficial flexor muscles, which include the **pronator teres (PT), flexor carpi radialis (FCR), palmaris longus (PL)** (sometimes absent), **flexor digitorum superficialis (FDS),** and **flexor carpi ulnaris (FCU)** (Table 6.2) (Fig. 6.8). At the wrist, note from lateral to medial, the **radial artery,** tendon of the FCR, **median nerve,** tendon of the PL, four tendons of the FDS, **ulnar artery** and **nerve,** and the tendon of the FCU. Palpate these structures at your own wrist.   A1. 55:15-59:07, 1:08:21-1:15:26/ G6.55/ C60/ R366/ N412-413, 416

In the cubital region, find the **brachial artery** and trace its terminal branches the **radial** and **ulnar arteries** to the wrist (Fig. 6.9). Small ulnar and radial recurrent arteries anastomose with vessels of the arm to form a collateral circulation around the elbow joint.   G6.42/ C64-67/ R395-397 N405, 417

2. Nerves. To fully expose the median nerve, cut the tendon of the PL 3 cm proximal to the wrist. Cut the FCR tendon 5 cm proximal to the wrist and detach the FDS from its origin on the radius. Trace the **median nerve** on the underside of the FDS and also trace and clean the ulnar nerve to the wrist. Proximally, the **ulnar nerve** passes deep to the two heads of the FCU. Find the **ulnar artery** and look for its **common interosseous branch.** Then trace the **anterior interosseous branch** of the common interosseous artery as it runs along the anterior side of the interosseous membrane.   A1. 1:11:20-1:15:20/ G6.57/ C64-67/ R398-399/ N417-418

Laterally at the elbow, identify the **brachioradialis** and open the furrow medial to this muscle to expose the **superficial branch of the radial nerve.** The **deep branch of the radial**

TABLE 6.2
MUSCLES ON ANTERIOR SURFACE OF FOREARM

Muscle	Proximal Attachment	Distal Attachment	Innervation	Main Actions
Pronator teres	Medial epicondyle of humerus and coronoid process of ulna	Middle of lateral surface of radius	Median n. (C6 and **C7**)	Pronates forearm and flexes elbow
Flexor carpi radialis	Medial epicondyle of humerus	Base of second metacarpal bone		Flexes wrist and abducts it
Palmaris longus	Medial epicondyle of humerus	Distal half of flexor retinaculum and palmar aponeurosis	Median n. (C7 and C8)	Flexes wrist and tightens palmar aponeurosis
Flexor carpi ulnaris	*Humeral head:* medial epicondyle of humerus *Ulnar head:* olecranon and posterior border of ulna	Pisiform bone, hook of hamate bone, and fifth metacarpal bone	Ulnar n. (C7 and **C8**)	Flexes wrist and adducts it
Flexor digitorum superficialis	*Humeroulnar head:* medial epicondyle of humerus, ulnar collateral ligament, and coronoid process of ulna *Radial head:* superior half of anterior border of radius	Bodies of middle phalanges of medial four digits	Median n. (C7, **C8,** and T1)	Flexes middle phalanges of medial four digits; acting more strongly, it flexes proximal phalanges and wrist
Flexor digitorum profundus	Proximal three-fourths of medial and anterior surfaces of ulna and interosseous membrane	Bases of distal phalanges of medial four digits	Medial part: ulnar n. (**C8** and T1) Lateral part: median n. (**C8** and T1)	Flexes distal phalanges of medial four digits; assists with flexion of wrist
Flexor pollicis longus	Anterior surface of radius and adjacent intercosseous membrane	Base of distal phalanx of thumb	Anterior interosseous n. from median (**C8** and T1)	Flexes phalanges of first digit (thumb)
Pronator quadratus	Distal fourth of anterior surface of ulna	Distal fourth of anterior surface of radius		Pronates forearm; deep fibers bind radius and ulna together

nerve pierces the **supinator muscle** to reach the posterior side of the forearm. (Fig. 6.9B).   **G6.57/ C63-64/ R397/ N417-418**

3. Deep Muscles. Identify and clean the three deep muscles which include the **flexor digitorum profundus, flexor pollicis**

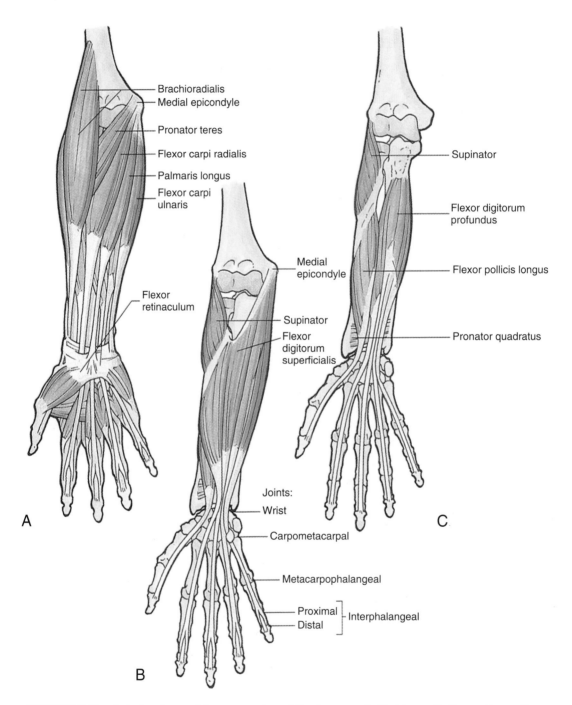

Brachioradialis
Medial epicondyle

Pronator teres

Flexor carpi radialis

Palmaris longus

Flexor carpi
ulnaris

Supinator

Flexor digitorum
profundus

Medial
epicondyle

Flexor
retinaculum

Flexor pollicis longus

Supinator
Flexor
digitorum
superficialis

Pronator quadratus

A

Joints:

Wrist

C

Carpometacarpal

Metacarpophalangeal

Proximal ⎱ Interphalangeal
Distal ⎰

B

FIGURE 6.8.   Anterior views of flexor muscles of the forearm. A. First layer. B. Second layer. C. Third and fourth layers.

longus, and **pronator quadratus** (Fig. 6.8). In your atlas and textbook, study the attachment of the superficial and deep flexor tendons to the second-fifth fingers, and review the innervation and actions of the forearm flexors and pronators (Fig. 6.8).   **G6.57-6.58/ C65/ R367-368/ N413, 417-418**

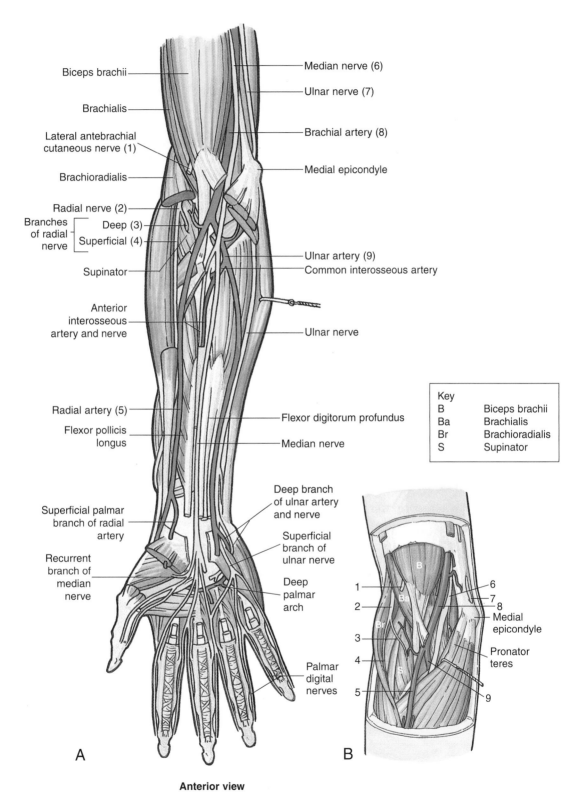

FIGURE 6.9. A. Forearm and hand. B. Cubital fossa. Numbers in B correspond to identifications in A.

Anterior view

IV. PALM

Learning Objectives

- Identify the key features of the carpals, metacarpals, and phalanges.
- Define which muscles contribute to the thenar and hypothenar eminences.
- Describe the motor and sensory innervation of the hand, and state where you could reliably test the sensory components of the radial, median, and ulnar nerves on the hand.
- On a cross-sectional diagram of the palm, label the muscles and tendons.
- Describe the attachments of the flexor tendons to a "typical" finger and their action.

Key Concepts

- Carpal tunnel
- Tendon sheaths and neurovascular bundles in the hand
- Autonomous zones of cutaneous innervation on the fingers

A. Introduction. The palm of the hand is flanked by two muscle masses: the thenar eminence (ball of the thumb) and the hypothenar eminence (ball of the little finger). Between these masses lies a superficial palmar aponeurosis and deep to it the tendons of the finger flexors and intrinsic hand muscles. **G6.61/ C80/ R3666-367/ N428**

Because of the sensory overlap in the hand, appreciate that radial sensory distribution is tested over the skin of the dorsal web space between the thumb and index finger. Median nerve sensation is tested on the palmar tip of the index finger and ulnar sensation on the palmar tip of the little finger.

B. Bony Landmarks

    1. On an articulated skeleton identify the following: **A1. 1:17:25-1:27:50/ G6.81/ C112-114/ R353-355/ N422-427**

        a. Eight carpal bones, including proximally the **scaphoid, lunate, triquetrum,** and **pisiform,** and distally the **trapezium, trapezoid, capitate,** and **hamate.**

        b. Five **metacarpal bones.**

        c. **Phalanges,** with a **proximal, middle** and **distal phalanx,** except for the thumb, which has only a proximal and distal phalanx.

C. Wrist

1. Carefully remove the **palmar aponeurosis.** Do not damage underlying vessels and nerves. Detach the **palmaris brevis,** reflect it medially, and note the **superficial palmar arterial arch.** Dissect several **digital arteries** arising from the arch (Fig. 6.10).   G6.63/ C88/ R398-400/ N428, 435

2. Starting at the forearm and wrist, push a probe deep to the flexor retinaculum and through the carpal tunnel. Cut through the **flexor retinaculum** to the probe and open up the **carpal tunnel** to visualize the nine muscle tendons and nerve. G6.59-6.61/ C82-83/ R367/N430-431

3. Dissect the median nerve into the palm and trace the small but important **recurrent branch of the median nerve** to the thenar muscles (Fig. 6.10B). This nerve innervates the intrinsic thenar muscles and may be lacerated in trauma involving the palm of the hand. Follow the digital branches of the median to the first three and one-half digits (note the median nerve sensory distribution on a dermatome illustration). G6.63/ C89/ R400/N429

D. Thenar Muscles

1. Identify and clean the **abductor pollicis brevis, opponens pollicis** and **flexor pollicis brevis** (Table 6.3) (Fig. 6.10) (save the recurrent nerve passing over its muscle belly!).   A1. 1:50:15-1:52:00/ G6.65/ C82-83/ R400/N429, 434

E. Hypothenar Muscles

1. Identify and clean the **abductor digiti quinti, opponens digiti quinti,** and **flexor digiti quinti** (sometimes absent) (quintiminimi).   A1. 1:52:00-1:53:38/ G6.67/ C82-83/ R400/ N429, 434

F. Digits

1. In the atlas, note the fibrous digital sheaths of the tendons passing into the fingers (Fig. 6.10). With the phalanges, these sheaths form an osseofibrous digital tunnel for the long flexor tendons. Appreciate that the tendons are surrounded by a common synovial sheath (of wrist and palm) and digital sheaths within the osseofibrous tunnels. Commonly, the first and fifth digital sheaths may communicate with the common synovial sheath. Sever the **osseofibrous tunnels of digits** 2-5 to expose the FDP tendons where they pass through the FDS tendons.   G6.66-6.69/ C82-84/ R399/ N430-432

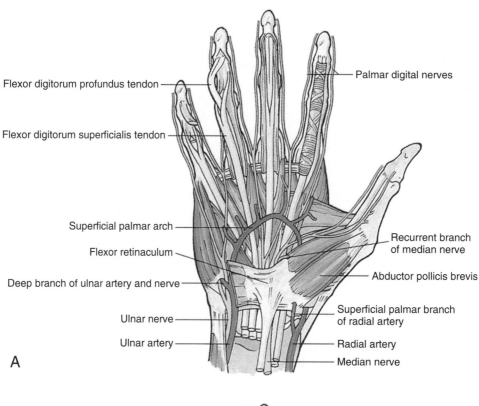

Flexor digitorum profundus tendon

Flexor digitorum superficialis tendon

Superficial palmar arch

Flexor retinaculum

Deep branch of ulnar artery and nerve

Ulnar nerve

Ulnar artery

Palmar digital nerves

Recurrent branch of median nerve

Abductor pollicis brevis

Superficial palmar branch of radial artery

Radial artery

Median nerve

A

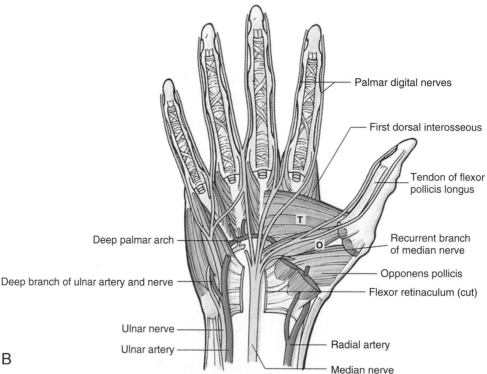

Palmar digital nerves

First dorsal interosseous

Tendon of flexor pollicis longus

Deep palmar arch

Recurrent branch of median nerve

Opponens pollicis

Flexor retinaculum (cut)

Deep branch of ulnar artery and nerve

Ulnar nerve

Ulnar artery

Radial artery

Median nerve

B

FIGURE 6.10.   A. Superficial dissection of palm, showing superficial palmar arch and distribution of median and ulnar nerves. B. Deeper dissection showing deep palmar arch and deep branch of ulnar nerve. T, transverse head of adductor pollicis; O, oblique head of adductor pollicis.

TABLE 6.3
MUSCLES OF HAND

Muscle	Proximal Attachment	Distal Attachment	Innervation	Main Actions
Abductor pollicis brevis	Flexor retinaculum and tubercles of scaphoid and trapezium bones	Lateral side of base of proximal phalanx of thumb	Recurrent branch of median n. (**C8** and T1)	Abducts thumb and helps oppose it
Flexor pollicis brevis	Flexor retinaculum and tubercle of trapezium bone			Flexes thumb
Opponens pollicis		Lateral side of first metacarpal bone		Opposes thumb toward center of palm and rotates it medially
Adductor pollicis	*Oblique head:* bases of second and third metacarpals, capitate, and adjacent carpal bones *Transverse head:* anterior surface of body of third metacarpal bone	Medial side of base of proximal phalanx of thumb	Deep branch of ulnar n. (C8 and **T1**)	Adducts thumb toward middle digit
Abductor digit minimi	Pisiform bone	Medial side of base of proximal phalanx of digit 5	Deep branch of ulnar n. (C8 and **T1**)	Abducts digit 5
Flexor digit minimi brevis	Hook of hamate bone and flexor retinaculum			Flexes proximal phalanx of digit 5
Opponens digiti minimi		Medial border of fifth metacarpal bone		Draws fifth metacarpal bone anteriorly and rotates it, bringing digit 5 into opposition with thumb
Lumbricals 1 and 2	Lateral two tendons of flexor digitorum profundus	Lateral sides of extensor expansions of digits 2–5	*Lumbricals 1 and 2:* median n. (C8 and **T1**)	Flex digits at metacarpophalangeal joints and extend interphalangeal joints
Lumbricals 3 and 4	Medial three tendons of flexor digitorum profundus		*Lumbricals 3 and 4:* deep branch of ulnar n. (C8 and **T1**)	
Dorsal interossei 1–4	Adjacent sides of two metacarpal bones	Extensor expansions and bases of proximal phalanges of digits 2–4	Deep branch of ulnar n. (C8 and **T1**)	Dorsal interossei abduct digits and palmar ones
Palmar interossei 1–3	Palmar surfaces of second, fourth, and fifth metacarpal bones	Extensor expansions of digits and bases of proximal phalanges of digits 2, 4, and 5		

2. Identify four small **lumbrical muscles** that pass from the FDP tendons on the radial side (lateral). These muscles insert into the extensor expansion on the dorsal side of the digits and flex the metacarpophalangeal joint (MP joint) while extending the interphalangeal joints. **A1. 1:46:08-1:50:03/ G6.65/ C83/ R400/ N431-433**

3. At the wrist, retract but do not cut the flexor tendons and expose the **pronator quadratus muscle** (Fig. 6.8). **G6.68/ C83/ R368, 406/ N418**

G. Deep Palmar Structures

1. Detach the flexor digiti quinti from the flexor retinaculum and follow the **deep branch of the ulnar nerve** across the palm (Fig. 6.10B). Cut muscles if necessary. Identify and clean the **adductor pollicis** and demonstrate the **deep palmar arterial arch**, if possible. Identify three **palmar interossei muscles** arising from the metacarpals of digits 2,4 and 5 and inserting into the proximal phalanges and extensor expansions. These muscles adduct the fingers (P<u>AD</u>) in relation to the middle finger, which is the reference axis for abduction and adduction in the hand. Dorsal interossei abduct the digits (D<u>AB</u>) (Table 6.3) **A1. 1:55:14-2:08:41/ G6.67/ C83-87/ R370-371, 404-406/ N434**

V. EXTENSOR FOREARM AND DORSAL HAND

Learning Objectives

- Describe the actions of the extensor muscles of the forearm and state how they are innervated.
- Label the functional muscle compartments of the forearm and name the vessels and nerves of each compartment on a cross-sectional diagram.
- Describe common nerve lesions of the upper extremity including Erb's palsy, Klumpke's paralysis, and lesions to the terminal branches of the brachial plexus.

Key Concepts

- Anatomical "snuff box"
- Supination
- Extensor expansion

A. Introduction. Extensors of the forearm are divided into a superficial and deep group. Superficial muscles generally arise from the lateral side of the elbow (lateral epicondyle) while the deep muscles arise mainly from the ulna and interosseous membrane. **G6.70/ C69-70/ R372/ N411, 419**

B. Anatomical "Snuff Box"

1. Identify the snuff box on yourself and feel the radial pulse in its fossa. On the cadaver, identify the tendons forming the snuff box. They are the **abductor pollicis longus** and **extensor pollicis brevis** anteriorly (anatomical position), and the **extensor pollicis longus** posteriorly (Table 6.4) (Fig. 6.11).

Within the snuff box, find the **radial artery** and trace it distally where it disappears between the two heads of the **first dorsal interosseous muscle** (your pinch muscle).   **A1. 1:40:49-1:43:26/ G6.71/ C75/ R370/ N436**

C. Superficial Extensors

1. On the dorsal wrist, identify the **extensor retinaculum** and the underlying tendons in their synovial sheaths (Fig. 6.12). Identify and clean the **brachioradialis, extensor carpi radialis longus,** and **extensor carpi radialis brevis.** More medially, clean the **extensor digitorum, extensor digiti minimi,** and **extensor carpi ulnaris.** Study the arrangement of the tendons as they pass through the extensor retinaculum (Fig. 6.12). Once you've thoroughly studied their arrangement, cut the retinaculum and reflect the muscle tendons medially to expose the deeper muscles. **A1. 1:38:57-1:40:39/ G6.67-6.71/ C69-71/ R372-373/ N411, 414**

D. Deep Extensors

1. Identify and clean the three muscles of the anatomical snuff box, the **abductor pollicis longus, extensor pollicis brevis,** and **extensor pollicis longus** (Fig. 6.11). Then clean the **supinator** and the **extensor indicis** (to index finger). Note that the **deep branch of the radial nerve** pierces the supinator and then runs along the interosseous membrane supplying innervation to the extensors (Fig. 6.11B). Finally, identify the four rather superficial lying **dorsal interosseous muscles.**   **G6.71/ C71/ R373/N410-411, 415, 434**

2. Extensor expansion. The distal ends of the extensor digitorum tendons become flattened to the thickness of a tendinous sheath, called the **extensor expansion** (Fig. 6.12A and C). The tendons of the lumbricals and interossei insert into this expansion, helping to extend the interphalangeal joints.

TABLE 6.4
MUSCLES ON POSTERIOR SURFACE OF FOREARM

Muscle	Proximal Attachment	Distal Attachment	Innervation	Main Actions
Brachioradialis	Proximal two-thirds of lateral supracondylar ridge of humerus	Lateral surface of distal end of radius	Radial n. (C5, **C6,** and C7)	Flexes elbow
Extensor carpi radialis longus	Lateral supracondylar ridge of humerus	Base of second metacarpal bone	Radial n. (C6 and C7)	Extend and abduct hand at wrist joint
Extensor carpi radialis brevis		Base of third metacarpal bone	Deep branch of radial n. (C7 and C8)	
Extensor digitorum	Lateral epicondyle of humerus	Extensor expansions of medial four digits	Posterior interosseous n. (**C7** and C8), a branch of the radial n.	Extends medial four digits at metacarpophalangeal joints; extends hand at wrist joint
Extensor digiti minimi		Extensor expansion of fifth digit		Extends fifth digit at metacarpophalangeal and interphalangeal joints
Extensor carpi ulnaris	Lateral epicondyle of humerus and posterior border of ulna	Base of fifth metacarpal bone		Extends and adducts hand at wrist joint
Anconeus	Lateral epicondyle humerus	Lateral surface of olecranon and superior part of posterior surface of ulna	Radial n. (C7, C8, and T1)	Assists triceps in extending elbow joint; stabilizes elbow joint; abducts ulna during pronation
Supinator	Lateral epicondyle of humerus, radial collateral and anular ligaments, supinator fossa, and crest of ulna	Lateral, posterior, and anterior surfaces of proximal third of radius	Deep branch of radial n. (C5 and **C6**)	Supinates forearm, i.e., rotates radius to turn palm anteriorly
Abductor pollicis longus	Posterior surfaces of ulna, radius, and interosseous membrane	Base of first metacarpal bone	Posterior interosseous n. (C7 and **C8**)	Abducts thumb and extends it at carpometacarpal joint
Extensor pollicis brevis	Posterior surface of radius and interosseous membrane	Base of proximal phalanx of thumb		Extends proximal phalanx of thumb at carpometacarpal joint
Extensor pollicis longus	Posterior surface of middle third of ulna and interosseous membrane	Base of distal phalanx of thumb		Extends distal phalanx of thumb at metacarpophalangeal and interphalangeal joints
Extensor indicis	Posterior surface of ulna and interosseous membrane	Extensor expansion of second digit		Extends second digit and helps to extend wrist

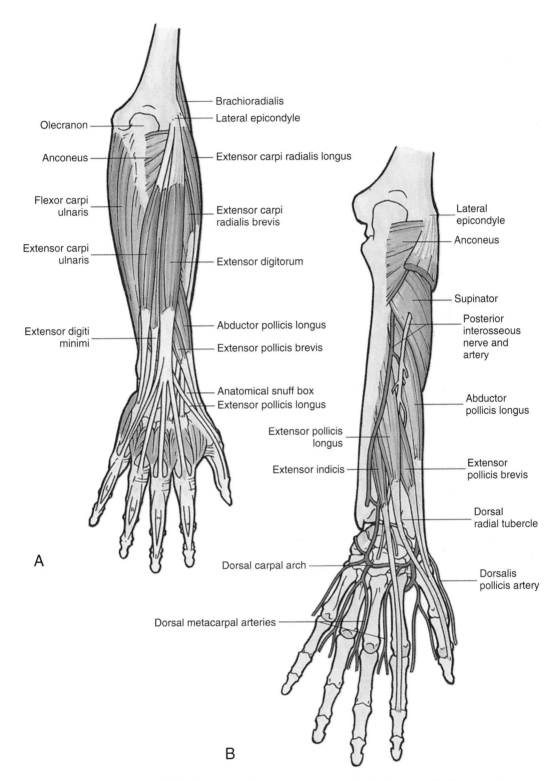

FIGURE 6.11. A. Superficial dissection of extensor muscles of the forearm. B. Deeper dissection of supinator and out-cropping muscles showing arteries and nerves.

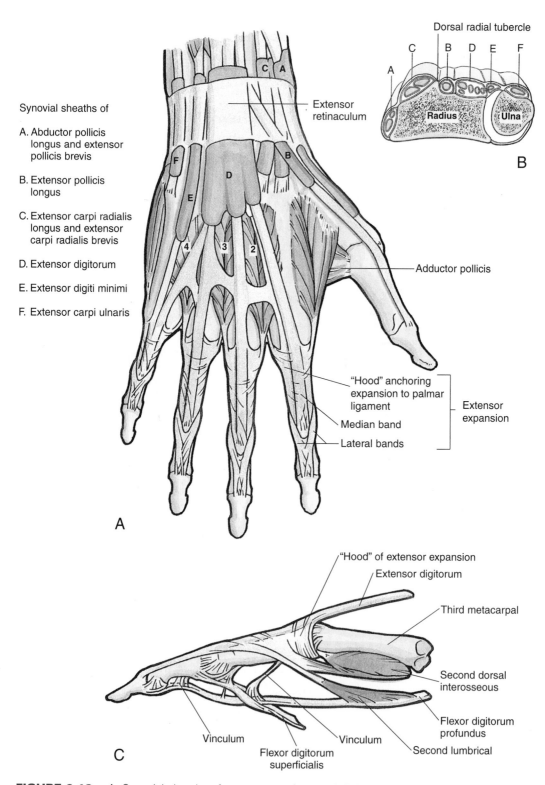

Synovial sheaths of

A. Abductor pollicis longus and extensor pollicis brevis

B. Extensor pollicis longus

C. Extensor carpi radialis longus and extensor carpi radialis brevis

D. Extensor digitorum

E. Extensor digiti minimi

F. Extensor carpi ulnaris

Extensor retinaculum

Dorsal radial tubercle

Radius   Ulna

B

Adductor pollicis

"Hood" anchoring expansion to palmar ligament

Median band

Lateral bands

Extensor expansion

A

"Hood" of extensor expansion

Extensor digitorum

Third metacarpal

Second dorsal interosseous

Flexor digitorum profundus

Second lumbrical

Vinculum

Vinculum

Flexor digitorum superficialis

C

FIGURE 6.12. A. Synovial sheaths of extensor tendons and digital extensor expansions; 1-4, dorsal interossei. B. Transverse section through distal radius and ulna showing tendons in their synovial sheaths. C. Lateral view of extensor expansion.

3. Review in your atlas and textbook cross-sections of the arm, forearm and hand, organizing your study into the muscle compartments, the function of the muscles in that compartment, their innervation and blood supply.   G6.33, 6.94, 6.96/ C131-138/ R402- 403/ N406, 419, 431

VI.   JOINTS OF THE UPPER LIMB

Learning Objectives

- Describe the importance of the rotator cuff in stabilization of the shoulder joint.
- Describe the action of the radius at the elbow during pronation and supination.
- Compare and contrast the articulations of the radius and ulna at the wrist joint.

Key Concepts

- Shoulder mobility and rotator cuff muscles
- Pronation and supination
- Radiocarpal joint and wrist fractures

A. Introduction. Carry out this dissection on only *one* of the upper limbs and preserve the other limb for review and study. Before beginning the dissection, review the bones comprising each joint.

B. Shoulder Joint

1. Remove all of the muscles of the anterior compartment of the arm around the shoulder and clean the insertion of the subscapularis.

This exposes the **fibrous capsule** of the joint on its anterior aspect (Fig. 6.13). Note that the front of the joint capsule is reinforced by the **glenohumeral ligaments.**   G6.37-6.39/ C96-101/ R356/ N394, 396

3. With the cadaver in the prone position, remove the posterior muscles from the shoulder joint capsule. Cut open the joint capsule, and with a probe explore the synovial cavity (Fig. 6.13).

4. With a saw or hammer and chisel, remove the head of the humerus close to the anatomical neck and identify the **gle-**

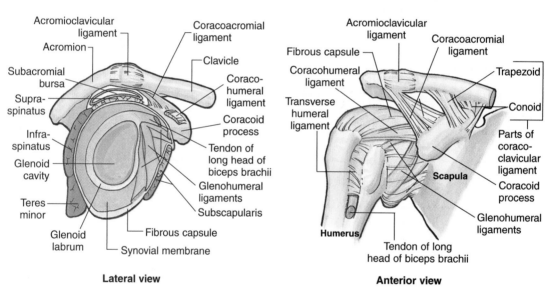

Tendon of long head of biceps brachii
Articular cartilage
Scapula
Fibrous capsule
Joint cavity
Humerus
Synovial membrane
Glenoid labrum
Fibrous capsule

Coronal section

Acromioclavicular ligament
Acromion
Subacromial bursa
Supra-spinatus
Infra-spinatus
Glenoid cavity
Teres minor
Glenoid labrum
Coracoacromial ligament
Clavicle
Coraco-humeral ligament
Coracoid process
Tendon of long head of biceps brachii
Glenohumeral ligaments
Subscapularis
Fibrous capsule
Synovial membrane

Lateral view

Acromioclavicular ligament
Fibrous capsule
Coracohumeral ligament
Transverse humeral ligament
Humerus
Coracoacromial ligament
Trapezoid
Conoid
Parts of coraco-clavicular ligament
Scapula
Coracoid process
Glenohumeral ligaments
Tendon of long head of biceps brachii

Anterior view

FIGURE 6.13.   Three views of the shoulder joint.

noid cavity, the **glenoid labrum,** and the **tendon of the long head of the biceps.** Define and clean the strong **coracoacromial ligament** (Fig. 6.13). Review the movements of the shoulder and the importance of the rotator cuff muscle tendons in stabilizing this joint.   **A1. 2:36-8:13/ G6.37/ C102/ R356/ N394**

C. Elbow Joint

1. Remove all of the soft tissues surrounding the elbow joint until the **ulnar collateral ligament** is visible medially. On the lateral side, clean the **radial collateral ligament** and note the

anular ligament, which encircles the head of the radius. Move the joint and observe the action of these ligaments. Note that the radius freely rotates in the anular ligament. Also note that with the hand pronated and elbow flexed, that the biceps brachii is a very strong supinator. Open the **joint capsule** anteriorly with a scalpel and explore the synovial capsule. Note thin synovial folds and fatpads intervening between the head of the radius and the capitulum of the humerus.   **A1. 52:54-54:45/ G6.51/ C116-118/ R357/ N408**

D. Wrist Joint. The wrist, or radiocarpal joint, is concerned with movements between the radius, and the scaphoid and lunate (Fig. 6.14). In your textbook, read about wrist fractures that commonly occur when someone falls on an out-stretched upper limb.

1. Remove all soft tissue around the wrist and on the palmar side notice the **radiocarpal ligaments** that hold the radius and

FIGURE 6.14. Top view shows the ligaments of the anterior wrist joint. Bottom view is a coronal section through the wrist joint.

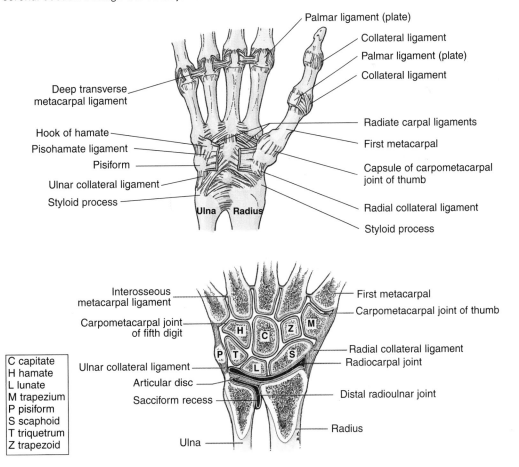

C capitate
H hamate
L lunate
M trapezium
P pisiform
S scaphoid
T triquetrum
Z trapezoid

carpals together. Place the hand in forced extension and cut through the radiocarpal ligaments to open the joint transversely. Examine the articulation and verify that an **articular disc** holds the distal ends of the radius and ulna together (Fig. 6.14). This disc articulates with the triquetrum when the wrist is adducted.   **A1. 1:17:25-1:33:54/ G6.83-6.88/ C126-130/ R358-359/ N424-425**

7

HEAD AND NECK

I. POSTERIOR TRIANGLE OF THE NECK

Learning Objectives

- Throughout Chapter 7, identify structures in bold print unless instructed to do otherwise.
- On a cross-sectional diagram, label the layers of cervical fascia.
- Describe the external jugular venous drainage of the neck.
- Describe the lymphatic drainage of the neck and identify the sites where "named" collections of lymph nodes would be found.
- List the major nerves of the cervical plexus and what they innervate.
- Identify the spinal accessory nerve and list what it innervates.
- Describe why pain from the diaphragm may sometimes be referred to the shoulder region.

Key Concepts

- Layers of cervical fascia dividing the neck into compartments
- Muscular triangles of the neck
- Cervical plexus

A. Introduction. We will begin the dissection of the head and neck region by exploring the muscular and visceral components of the neck first. Then we will proceed onto the face, remove the brain, **151**

study the orbital region, and then work our way systematically down through the head toward the deep structures of the neck (pharynx and larynx).

1. First, notice that the neck is bounded by three layers of deep fascia, which separate the neck into compartments. These fascia layers are the **investing fascia, pretracheal fascia,** and the **prevertebral fascia.** These layers form natural planes for surgical dissection, limit the spread of infections, and provide smooth, mobile surfaces for muscle contraction and movements of the visceral neck during swallowing. Read about them in your textbook.   **G8.2/ C704-705/ R150-152**

2. The boundaries of the posterior triangle are the **sternocleidomastoid muscle,** the **trapezius** and the **middle third of the clavicle.**   **G8.3/ C693/ R153/ N362**

FIGURE 7.1.   Dashed lines show skin incisions.

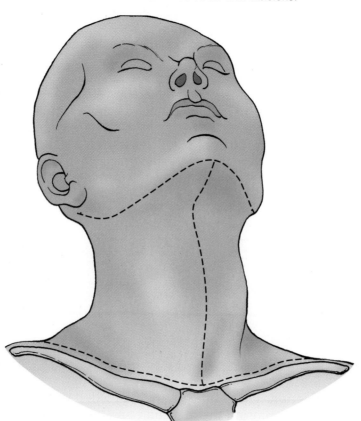

B. Superficial Structures

1. Make the skin incisions shown in Figure 7.1 and reflect the skin posteriorly to well behind the ear. Note the underlying **platysma muscle,** a muscle of facial expression, which has migrated onto the neck. Beneath the platysma lie the **supraclavicular cutaneous nerves** (C3-4) (medial, intermediate and lateral). Slightly superior to the middle of the posterior border of the sternocleidomastoid muscle, locate the **spinal accessory nerve** (11th cranial nerve, CN. XI; CN and a Roman numeral identify the cranial nerves, of which there are 12 pairs) coursing downward toward the trapezius muscle (Fig. 7.2). Using your scissors, incise and spread the tough fascial covering of the posterior triangle and locate the **lesser occipital nerve** (C2-3) emerging close to CN. XI. From Figure 7.2, note the direction that each nerve takes as it traverses the posterior triangle. **A4. 26:40-34:30, A5. 1:21:05-1:23:30/ G8.5-8.6/ C696-699/ R61, 170, 172/ N21, 27**

FIGURE 7.2. Posterior cervical triangle.

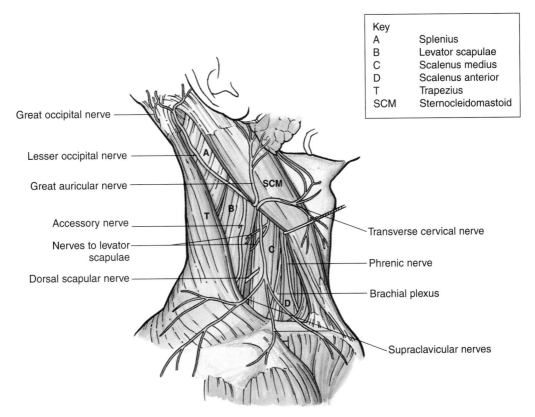

Key
A — Splenius
B — Levator scapulae
C — Scalenus medius
D — Scalenus anterior
T — Trapezius
SCM — Sternocleidomastoid

Great occipital nerve

Lesser occipital nerve

Great auricular nerve

Accessory nerve

Nerves to levator scapulae

Dorsal scapular nerve

SCM

Transverse cervical nerve

Phrenic nerve

Brachial plexus

Supraclavicular nerves

2. Next locate the **great auricular nerve** (C2-3) which ascends posterior and parallel with the external jugular vein on the sternocleidomastoid. Try to identify the small transverse cervical nerve (C2-3) supplying skin over the anterior neck.

C. Removal of the Clavicle

1. Remove the middle portion of the clavicle by cutting it laterally with a saw close to the attachments of the deltoid and trapezius muscles, and medially where the sternocleidomastoid muscle attaches. With the clavicle removed, look for the slender subclavius muscle and remove it as well. Examine the **omohyoid** (omo refers to shoulder) and its superior and inferior bellies. Its intermediate tendon is tethered by a fascial sling to the clavicle.   **G8.7/ C700-701/ R171/ N26-28**

2. Locate the **subclavian vein** (follow the external jugular vein until it drains into this vein) (see Fig. 7.4). Also, identify two arteries: the **transverse cervical artery** which runs about 2-3 cm superior to the clavicle and deep to the omohyoid muscle, on its way towards the levator scapulae (Fig. 7.3) and the **suprascapular artery** which is headed toward

FIGURE 7.3.   View of the brachial plexus, subclavian vessels, and muscles in the lower neck. B, levator scapulae; C, scalenus medius; D, scalenus anterior; E, scalenus posterior; T, trapezius; SCM, sternocleidomastoid.

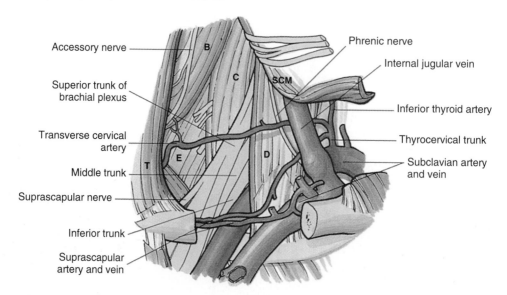

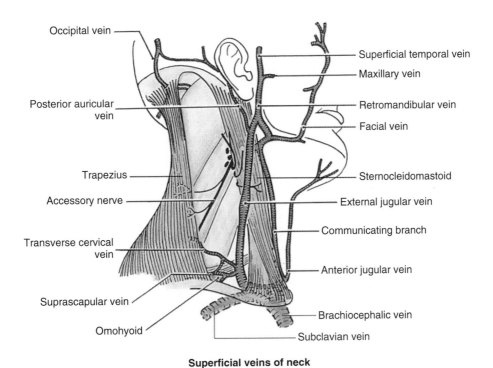

Superficial veins of neck

FIGURE 7.4. Superficial veins of the neck.

the suprascapular notch of the scapula. Both arteries generally pass anterior to the scalenus anterior muscle. Note: these arteries can be variable in their origin and locations, but usually arise from the thyrocervical trunk of the subclavian.

D. Deep Structures

1. Remove the fascia floor of the triangle (prevertebral fascia) to reveal the **splenius capitis, levator scapulae, scalenus posterior, scalenus medius,** and **scalenus anterior muscles** (Table 7.1). The scalenus anterior and medius attach to the first rib and in the interval between the two muscles one finds the **subclavian artery** and portions of the **brachial plexus** (nerve plexus to upper limb) (Fig. 7.3). Identify and preserve the **phrenic nerve** (C3-5) on the anterior surface of the anterior scalene muscle.   **G8.8/ C701-703/ R178/ N25-28, 173, 175**

2. Nerves piercing the middle scalene are motor branches to the rhomboids and serratus anterior muscles. These structures are included in the dissection of the upper limb and brachial plexus.

TABLE 7.1
MUSCLES IN LATERAL ASPECT OF NECK

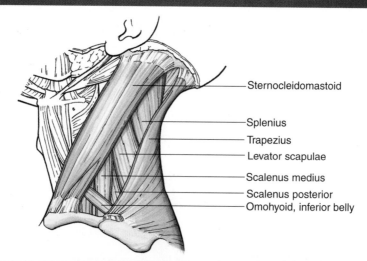

- Sternocleidomastoid
- Splenius
- Trapezius
- Levator scapulae
- Scalenus medius
- Scalenus posterior
- Omohyoid, inferior belly

Muscle	Origin	Insertion	Innervation	Actions
Platysma	Inferior border of mandible, skin and subcutaneous tissues of lower face	Fascia covering superior parts of pectoralis major and deltoid muscles	Cervical branch of facial n. (CN VII)	Draws corners inferiorly and widens mouth as in expressions of sadness and fright; draws the skin of neck superiorly
Trapezius	Medial third of superior nuchal line, external occipital protuberance, ligamentum nuchae, and spinous processes of C7–T12 vertebrae	Lateral third of clavicle, acromion and spine of scapula	Spinal cord of accessory in (CN XI) and C3 and C4 nn.	Elevates, retracts, and rotates scapula
Sternocleidomastoid	Sternal head: anterior surface of manubrium of sternum Clavicular head: superior surface of medial third of clavicle	Lateral surface of mastoid process of temporal bone and lateral half of superior nuchal line	Spinal root of accessory n. (CN XI) and C2 and C3 nn.	Tilts head to one side, i.e., laterally; flexes neck (cervical vertebrae) and rotates it so face is turned superiorly toward opposite side; acting together, the two muscles flex the neck
Splenius capitis	Inferior half of ligamentum nuchae and spinous processes of superior six thoracic vertebrae	Lateral aspect of mastoid process and lateral third of superior nuchal line	Dorsal rami of middle cervical spinal nn.	Laterally flees and rotates head and neck to same side; acting bilaterally, they extend head and neck
Levator scapulae	Posterior tubercles of transverse processes of C1–C4 vertebrae	Superior part of medial border of scapula	Dorsal scapular n. (C5) and cervical spinal nn. (C3 and C4)	Elevates scapula and tilts its glenoid cavity inferiorly by rotating scapula
Scalenus posterior	Posterior tubercles of transverse processes of C4–C6 vertebrae	External border of 2nd rib	Ventral rami of cervical spinal nn. (C7 and C8)	Flexes neck laterally; elevates 2nd rib during forced inspiration
Scalenus medius	Posterior tubercles of transverse processes of C2–C7 vertebrae	Superior surface of 1st rib, posterior to groove for subclavian a.	Ventral rami of cervical spinal nn.	Flexes neck laterally; elevates 1st rib during forced inspiration

II. ANTERIOR TRIANGLE AND ROOT OF THE NECK

Learning Objectives

- List the functions of the infrahyoid ("strap") muscles.
- List the nerves found in the carotid triangle and describe what they innervate.
- Compare and contrast the blood supply of the external and internal carotid arteries.
- List the structures found in the carotid sheath.
- Identify the important functions of the thyroid and parathyroid glands, and describe their blood supply.

Key Concepts

- Muscular triangles
- Carotid sheath
- Ansa cervicalis

A. Introduction

1. First, study the bones and cartilages of the neck: A4. 00:55-34:30, A4. 1:38:55-1:45:15, A4. 2:07:10-2:10:30/ G8.46/ C712/ R154/ N24, 71

   a. **Hyoid bone.** Palpate your own hyoid bone (the body) just above the thyroid cartilage.

   b. **Thyroid cartilage.** Palpate the laryngeal prominence (Adam's apple).

   c. **Cricoid cartilage.** Lies at level of C6 vertebra.

   d. **Trachea.**

B. Superficial Structures

1. Look for the **facial vein, retromandibular vein** and, if present, the small **anterior jugular vein,** and review the external jugular system (Fig. 7.4).   G8.11/ C709/ R78/ N26

C. Muscular Triangle

1. This triangle includes the "strap" muscles that lie anterior to the trachea. The superficial layer of strap muscles consists of the **superior belly of the omohyoid** and **sternohyoid.** Deep to these are the **sternothyroid** and short **thyrohyoid muscles** (Table 7.2). Spread the infrahyoid muscles apart and identify the **cricothyroid membrane** stretching between the **thyroid** and **cricoid cartilages.**   G8.12-8.13/ C707/ R171/ N23-26

TABLE 7.2
SUPRAHYOID AND INFRAHYOID MUSCLES

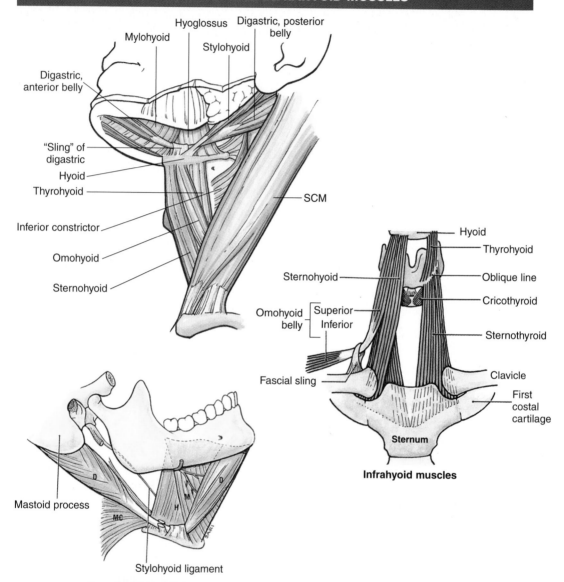

Suprahyoid muscles

Infrahyoid muscles

Muscle	Origin	Insertion	Innervation	Actions
Mylohyoid	Mylohyoid line of mandible	Raphe and body of hyoid bone	Mylohyoid n., a branch of inferior alveolar n.	Elevates hyoid bone, floor of mouth, and tongue during swallowing and speaking
Geniohyoid	Inferior mental spine of mandible	Body of hyoid bone	C1 via the hypoglossal n. (CN XII)	Pulls hyoid bone anterosuperiorly, shortens floor of mouth, and widens pharynx
Stylohyoid	Styloid process of temporal bone	Body of hyoid bone	Stylohyoid branch of facial n. (CN VII)	Elevates and retracts hyoid bone, thereby elongating floor of mouth

			TABLE 7.2 (Continued)	
		SUPRAHYOID AND INFRAHYOID MUSCLES		
Muscle	Origin	Insertion	Innervation	Actions
Digastric	*Anterior belly:* digastric fossa of mandible *Posterior belly:* mastoid notch of temporal bone	Intermediate tendon to body and superior (greater) horn of hyoid bone	*Anterior belly:* mylohyoid n., a branch of inferior alveolar n.; *Posterior belly:* facial n. (CN VII)	Depresses mandible; raises hyoid bone and steadies it during swallowing and speaking
Sternohyoid	Manubrium of sternum and medial end of clavicle	Body of hyoid bone	C1–C3 from ansa cervicalis	Depresses hyoid bone after it has been elevated during swallowing
Sternothyroid	Posterior surface of manubrium of sternum	Oblique line of thyroid cartilage	C2 and C3 by a branch of ansa cervicalis	Depresses hyoid bone and larynx
Thyrohyoid	Oblique line of thyroid cartilage	Inferior border of body and superior (greater) horn of hyoid bone	C1 via hypoglossal n. (CN XII)	Depresses hyoid bone and elevates larynx
Omohyoid	Superior border of scapula near suprascapular notch	Inferior border of hyoid bone	C1–C3 by a branch of ansa cervicalis	Depresses, retracts, and steadies hyoid bone

D. Carotid Triangle

1. This triangle is bound by the **superior belly of the omohyoid, posterior belly of the digastric,** and **anterior border of the sternocleidomastoid.** Divide the sternocleidomastoid muscle about 5 cm above its inferior attachments and reflect the muscle toward the mastoid process while preserving the spinal accessory nerve. Cut the facial vein where it empties into the internal jugular vein.

2. **Nerves.** Palpate and locate the tip of the greater horn of the hyoid bone. Just superior to the tip, find the **hypoglossal nerve (CN. XII)** where it crosses the carotid sheath anteriorly and laterally. Now try to find the **superior root of the ansa cervicalis,** which is composed mainly of fibers from C1 that run with the CN. XII. The **inferior root** (C2-3) descends from the more posterior superior neck region to join the superior root, together forming a loop overlying the carotid sheath. The ansa innervates the infrahyoid muscles and often is enmeshed in the carotid sheath.   **G8.13/ C701-702/ R169, 176/ N27, 29**

Next, find the **vagus nerve** (CN. X) by carefully opening the carotid sheath, if not already done. It lies within the carotid

sheath between the **common carotid artery** and **internal jugular vein.** Look for the **superior laryngeal nerve** of CN. X and its larger branch, the **internal laryngeal nerve.** To find this nerve, relax the neck, and then sever the omohyoid, sternohyoid, and thyrohyoid muscles close to the hyoid bone. This exposes the **thyrohyoid membrane** and the internal laryngeal nerve can be seen piercing this membrane. This nerve is sensory to the larynx above the vocal cords. The other portion of the superior laryngeal nerve is its very small **external laryngeal nerve** which innervates the **cricothyroid muscle.**
G8.26/ C702-703/ R176/ N28-29, 65, 74

3. **Arteries.** Remove the remainder of the carotid sheath and identify the **common carotid artery, internal carotid artery** and the closely applied **internal jugular vein.** Identify the **external carotid artery** and its first five branches (Fig. 7.5): **A5.**
1:25:48-1:28:11, A5. 1:21:05-1:23:30/ G8.17/ C746/ R178/ N27, 29, 63

a. **Superior thyroid.** Supplies the upper part of the thyroid gland and gives off the **superior laryngeal artery,** which pierces the thyrohyoid membrane with the internal laryngeal nerve.

b. **Lingual.** To the tongue.

c. **Facial.** Arises just above the lingual and, in 20% of all cases, the lingual and facial form a common trunk.

d. **Occipital.** Provides a muscular branch to the sternocleidomastoid muscle.

FIGURE 7.5. Subclavian and carotid arteries.

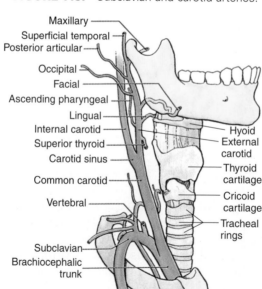

e. **Ascending pharyngeal.** Arises close to the carotid bifurcation and is difficult to find since it is deeply placed.

Clean the carotid bifurcation and note the dilated proximal portion of the internal carotid artery. This is the **carotid sinus region** (specialized baroreceptor which monitors blood pressure changes). In the bifurcation, closely adherent to the internal carotid artery is the carotid body, another specialized receptor (chemoreceptor) which monitors blood O_2 and CO_2 levels, and pH (innervated by a small branch of CN. IX).

4. **Veins.** Identify the **internal jugular vein, retromandibular vein, common facial vein, lingual vein** and **superior thyroid vein** (Fig. 7.4). **G8.11/ C747/ R172-173/ N26**

E. Submandibular Triangle

1. Also known as the digastric triangle, this region is bounded by the mandible and the **anterior** and **posterior bellies of the digastric muscle.**

2. Identify the **submandibular salivary gland.** It is wrapped around the posterior free border of the **mylohyoid muscle.** Remove the superficial portion of the gland but preserve the facial artery and vein. Define and clean the **anterior** and **posterior bellies of the digastric muscle.** The two bellies are connected by an intermediate tendon, which attaches to the hyoid bone. The slender **stylohyoid muscle** splits to surround this tendon. **G8.16-8.17/ C724-727/ R177/ N23, 27, 47, 55**

3. Pull the anterior belly of the digastric medially to find the **mylohyoid nerve,** sheltered by the inferior border of the mandible. This nerve is a branch of the trigeminal nerve.

F. Thyroid Gland

1. Expose the gland and verify that it consists of **right** and **left lobes** and an intervening **isthmus** (Fig. 7.6). Sometimes, a **pyramidal lobe** is found ascending from the isthmus. **A4. 2:35:30-2:38:20/ G8.26/ C710-712/ R178-181/ N68**

2. Examine the gland's blood supply: **superior** and **inferior thyroid arteries,** and three veins (superior, middle and inferior) (Fig. 7.6C). The inferior thyroid artery often is looped and is a branch of the thyrocervical trunk of the subclavian artery.

3. Now cut the isthmus of the gland, turn the lobes laterally and probe for the **recurrent laryngeal nerves** that ascend on each

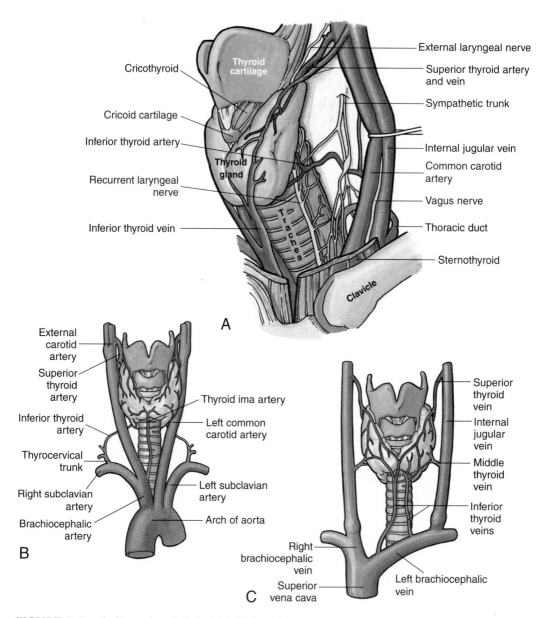

FIGURE 7.6. A, Root of neck (left side). B, Arterial supply. C, Venous drainage.

side posterior to the gland and often lie in the groove between
the **trachea** and **esophagus** (Fig. 7.6). These nerves innervate
the muscles associated with the vocal cords, are sensory to the
larynx below the cords, and are branches of the vagus nerve.
G8.27/ C715/ R179/ N65, 68-70

G. Parathyroid Glands

1. On the left side only, cut all blood vessels to the thyroid gland,
enucleate the left lobe and look on its posterior aspect for pea-

sized **parathyroid glands** (usually a superior and inferior gland, although variable numbers exist and may be found even inferiorly in the thymus gland!). These small glands are difficult to discern in the cadaver because of the fixation process.   G8.36/ C713/ R163/ N69-70

H. Base of the Neck

1. Cut the common carotid artery and internal jugular vein on the left side (but not the vagus!) 2 cm superior to the clavicle, and reflect them superiorly. Look for the **thoracic duct,** which enters the angle between the **left internal jugular vein** and **left subclavian vein** (Fig. 7.6).   G8.32/ C715/ R169/N66, 197, 227

2. Find the **transverse cervical** and **suprascapular arteries** again and trace them back to their origin from the **thyrocervical trunk** of the **subclavian artery** (Fig. 7.3). This trunk lies near the medial border of the anterior scalene muscle. The third branch of this trunk is the **inferior thyroid artery** (Fig. 7.6B). Variations are common in this area.   G8.30/ C702-703/ R178/N28, 214, 228, 398

3. Next, find the **vertebral artery,** the first and largest branch of the subclavian. This artery usually passes through the transverse foramen of C6.   G8.14/ C719-723/R158/ N14, 28, 130

4. Finally, identify the **sympathetic trunk** and its **chain ganglia** posterior to the carotid sheath.   G8.30/ C703/ R178/ N65

III. FACE

Learning Objectives

- Begin to examine the bony skull and identify its anatomical features.
- Identify several of the more important muscles of facial expression, especially those of the forehead, around the eye and mouth, and the chin.
- List the five terminal branches of the facial nerve. B C T M Z   skin
- Name the layers of the scalp and describe why scalp wounds bleed profusely.   C.T. aponeurosis Loose CT
- List the three divisions of the trigeminal nerve.   mandibular maxillary ophthalmic   periosteum

Key Concepts

- Muscles of facial expression
- Layers of scalp
- Cutaneous innervation of face via CN. V

A. Bony Landmarks

1. Begin your study of the skull. Do not mark it with a pen or pencil, and be especially careful of the orbital region as some of the bones are paper thin. Initially, learn the following bones and features and then add to your knowledge as we dissect various regions of the head. **A4. 35:30-55:45/ G7.1-7.2/ C752-759/ R25-26/ N1-5**

    a. **Frontal bone**

    b. **Maxilla**

    c. **Zygomatic bone**

    d. **Mandible**

    e. **Teeth.** In the adult, 32 teeth with 16 in each jaw. Each half of the jaw contains 2 incisors, 1 canine, 2 premolars, and 3 molars. The third molar is known as the "wisdom tooth."

    f. **Lacrimal bone**

    g. **Coronal suture.** Separates frontal and **parietal bones.**

    h. **Sagittal suture.** Separates the two parietal bones.

    i. **Bregma.** Meeting point of coronal and sagittal sutures.

    j. **Occipital bone**

    k. **Lambdoid sutures.** Separate occipital from parietal bones.

    l. **Lambda.** Meeting point of lambdoid and sagittal sutures.

    m. **External auditory meatus,** situated in the **temporal bone.**

    n. **Zygomatic arch**

    o. **Mastoid process**

    p. **Styloid process**

    q. **Stylomastoid foramen**

B. Facial Nerve and Vessels

1. Make the skin incisions shown in Figure 7.7. In some places, the skin is quite thin so be careful with your dissection. Beneath the skin you will encounter the muscles of facial expression which insert into the skin and are innervated by the facial nerve (CN. VII) (Table 7.3). Look for representative branches of the **facial nerve** which collectively innervate facial muscles in five regions: **temporal, zygomatic, buccal, mandibular,** and **cervical** (Fig. 7.8). The facial nerve gains access to the face by exiting the stylomastoid foramen, passing through the **parotid gland** and then "fanning out" over the lateral aspect of the face. Over the cheek region, be sure to identify and preserve the **parotid duct** where it crosses the **masseter muscle.** **G7.10/ C728-730/ R76/ N19**

[handwritten notes in left margin:] Facial nerve — 5 branches — 1. temporal — 2. zygomatic — 3. buccal — 4. mandibular — 5. cervical

[handwritten note at bottom:] * stylomastoid framen

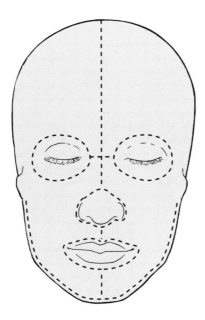

FIGURE 7.7. Dashed lines show skin incisions.

TABLE 7.3
MAIN MUSCLES OF FACE[a]

Muscle	Origin	Insertion	Action(s)
Frontalis	Epicranial aponeurosis	Skin of forehead	Elevates eyebrows and forehead
Orbicularis oculi	Medial orbital margin, medial palpebral ligament, and lacrimal bone	Skin around margin of orbit; tarsal plate	Closes eyelids
Nasalis	Superior part of canine ridge of maxilla	Nasal cartilages	Draws ala (side) of nose toward nasal septum
Orbicularis oris	Some fibers arise near median plane of maxilla superiorly and mandible inferiorly; other fibers arise from deep surface of skin	Mucous membrane of lips	Compresses and protrudes lips (e.g., purses them during whistling and sucking)
Levator labii superioris	Frontal process of maxilla and infraorbital region	Skin of upper lip and alar cartilage of nose	Elevates lip, dilates nostril, and raises angle of mouth
Platysma	Superficial fascia of deltoid and pectoral regions	Mandible, skin of cheek, angle of mouth, and orbicularis oris	Depresses mandible and tenses skin of lower face and neck
Mentalis	Incisive fossa of mandible	Skin of chin	Elevates and protrudes lower lip
Buccinator	Mandible, pterygomandibular raphe, and alveolar processes of maxilla and mandible	Angle of mouth	Presses cheek against molar teeth, thereby aiding chewing; expels air from oral cavity as occurs when playing a wind instrument

[a]All these muscles are supplied by the facial nerve (CN VII).

galea aponeurotica

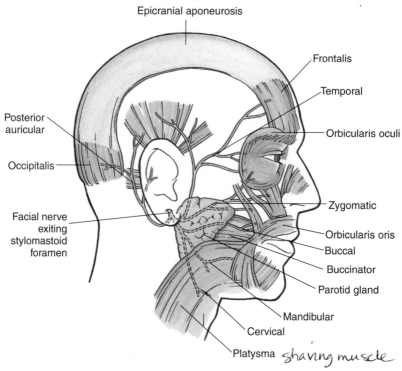

Epicranial aponeurosis

Frontalis

Temporal

Orbicularis oculi

Posterior
auricular

Occipitalis

Zygomatic

Orbicularis oris

Facial nerve
exiting
stylomastoid
foramen

Buccal

Buccinator

Parotid gland

Mandibular

Cervical

Platysma *shaving muscle*

FIGURE 7.8.   Terminal branches of facial nerve and some of the muscles of facial
expression.

2. Define the anterior border of the masseter, look for the buccal
fatpad and remove this fatpad to expose the **buccinator muscle,** which is pierced by the parotid duct. Locate the **facial artery** and **vein** where they cross the mandible and see if you
can detect your own facial artery pulse at this point.   **G7.10/
C735/ R77/ N17-19**

C. Muscles of the Mouth

1. Identify the **depressor anguli oris** ("sad" muscle), the **zygo-
maticus major** and **levator labii superioris** ("happy" muscles) (Fig. 7.8). The important sphincter muscle of the mouth
that allows us to purse our lips and kiss is the **orbicularis oris.**
(Outside of lab, look in a mirror, pull your lower lip downward and try to identify labial glands beneath the mucous
membrane just inside the lip. These glands, and the salivary
glands, help keep the lips moist.) At the red line of the cadaver's lower lip, incise the orbicularis oris muscle and iden-

tify the **inferior labial artery,** a branch of the facial artery.
A5. 1:13-5:59/ G7.13/ C728/ R60/ N17, 20-21

D. Inspection of the Eyelids

1. Carefully remove the skin on the eyelids and identify the **orbicularis oculi,** which consists of two parts: a thick orbital portion surrounding the orbital margin (for closing our eyes tightly), and a thin palpebral portion contained in the eyelids and used in blinking.   **G7.9/ C785-786/ R60/ N20-21**

E. Scalp

1. The scalp consists of the Skin, Connective tissue (superficial fascia), Aponeurosis comprised of the galea aponeurotica (connecting the frontalis and occipitalis muscles), Loose areolar connective tissue, and the Pericranium (outer layer of the bony skull). **SCALP.**   **A5. 6:00-9:14/ G7.12/ C729/ R85/N17, 21, 94**

2. Identify the **frontalis** and **occipitalis muscles,** and the **galea aponeurotica** which unites them. Most of the blood vessels and nerves of the scalp run in the connective tissue layer (superficial fascia) which is quite tough. When lacerated, scalp vessels bleed profusely because they are held open by the dense fibrous tissue of the second layer of the scalp and cannot retract. Note on your own scalp that the skin "glides" over the bony skull because it moves easily over the loose areolar connective tissue layer that lies just above the pericranium.

F. Sensory Cutaneous Nerves

1. From your atlas and textbook review the sensory distribution of the face by the three divisions of the trigeminal nerve (CN. V). Now locate the **supraorbital nerves** (from the ophthalmic division of CN. V) where they exit the skull via the **supraorbital foramen** (or notch). Probe through the orbicularis oculi to find this nerve. Next locate the **infraorbital nerve** branches (maxillary division of CN. V), which exit the skull through the **infraorbital foramen.** You may need to reflect the levator labii superioris to find the nerve.

Finally, probe through the muscle inferior to the lower lip and find the **mental nerve** (mandibular division of CN. V) exiting from the **mental foramen.**   **G7.13/ C785/ R82/ N18**

IV. INTERIOR OF SKULL AND BRAIN REMOVAL

Learning Objectives

- List the meningeal layers, and describe the blood flow from any dural venous sinus back to the heart.
- Identify the major lobes of the brain.
- On a diagram, label the major arteries on the base of the brain and the Circle of Willis.
- Identify each cranial nerve on the brain.
- Describe why the cavernous sinus is important.
- Compare and contrast the various types of intracranial hematomas.

Key Concepts

- Meninges of the CNS
- Dural venous sinuses
- Circle of Willis
- Ventricular system and cerebrospinal fluid
- Cranial nerves pass through foramen

A. Introduction. This dissection may require two lab periods, depending on your individual schedules. Check with your instructor.

    1. On a dried skull note that the **calvaria** (skull cap) consists of three parts: a compact outer lamina, a compact and very hard inner lamina, and the diploe, a layer of spongy bone between the two compact lamina. Numerous diploic veins fill this space and communicate with vessels of the scalp and vessels of the dura mater covering the brain.   **G7.16/ C764/ R34/ N4**

B. Removal of the Calvaria

    1. With your scalpel, incise the temporalis fascia outlining the **temporalis muscle** and reflect the muscle inferiorly to the level of the zygomatic arch. Scrape the bony skull clean and place a rubber band around the skull's circumference. Anteriorly, be sure the band is at least 2 cm above the supraorbital margin and posteriorly, 2 cm above the external occipital protuberance (inion). Using the band as a guide, draw a circle around the skull with a pencil, remove the band, and with a saw cut through the external lamina along the line (Fig. 7.9). Turn the body several times until your cut completely encircles the calvaria. Cut through the diploe and just into the inner lamina. Then, with a mallet and chisel, break the inner lamina and pry the calvaria open with the inserted chisel. Complete removal of

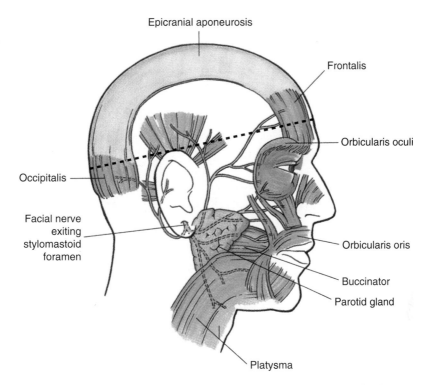

Epicranial aponeurosis

Frontalis

Orbicularis oculi

Occipitalis

Facial nerve
exiting
stylomastoid
foramen

Orbicularis oris

Buccinator

Parotid gland

Platysma

FIGURE 7.9. Dashed line shows saw cut to remove the calvaria.

the calvaria will require you to detach the inner lamina of the skull from the underlying dura which is closely applied to the bone. Keep the dura on the brain's surface, if possible.

2. Next, place the cadaver in the prone position and detach all muscles that lie deep within the back of the neck that attach to the occipital bone. Incise the **posterior atlanto-occipital membrane** transversely extending your cut from vertebral artery to vertebral artery. Scrape the bone clean and with a hand saw remove the wedge of bone as indicated in Figure 7.10 (cut from A to B). Be sure your cut passes through the **foramen magnum** posteriorly.

C. Brain Meninges

1. **Dura Mater.** Note that the brain is covered by the **dura mater,** which consists of two layers: a rough outer layer adherent to the skull where it forms the endocranium, and a smooth inner layer. Normally, these two layers are indistinguishable except where they separate to enclose the dural venous sinuses. **A5. 11:34-19:28/ G7.19/ C766/ R85-88/N95**

2. Identify the **middle meningeal artery** and its branches which supply the dura and overlying skull. Look for small nutrient

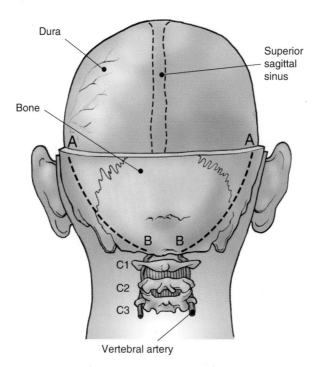

FIGURE 7.10. Dashed lines show saw cuts to remove bone wedge.

foramina on the calvaria which convey blood vessels into the bone.

3. Identify the **superior sagittal dural sinus** and right and left **transverse dural sinuses** (Fig. 7.11). Understand the importance of the dural sinuses in returning blood to the internal jugular vein and heart. Along the superior sagittal sinus look for **lacunae laterales** and the **arachnoid granulations.** **G7.21/ C766-768/ R87-88/N94-96**

4. Make incisions in the dura on each side which correspond to the coronal sutures. Try not to incise the underlying arachnoid. Also, cut the dura parallel to the superior sagittal sinus (see dashed lines in dura in Figure 7.10) and transverse sinuses. Keep your cuts about 2 cm lateral to the superior sagittal sinus and above the transverse sinuses. Now, expose the **cerebral hemispheres** by reflecting the dural flaps. Then cut the dura inferior to the transverse sinuses and along the margins of the removed bony wedge. This exposes the **cerebellar hemispheres.**

5. **Arachnoid and Pia Mater.** Study the **arachnoid** and, on one part of the cerebral hemisphere, remove the arachnoid to expose the **pia mater,** which encases the brain tissue even between the sulci and fissures. Follow several cerebral veins to their entry into the superior sagittal sinus.

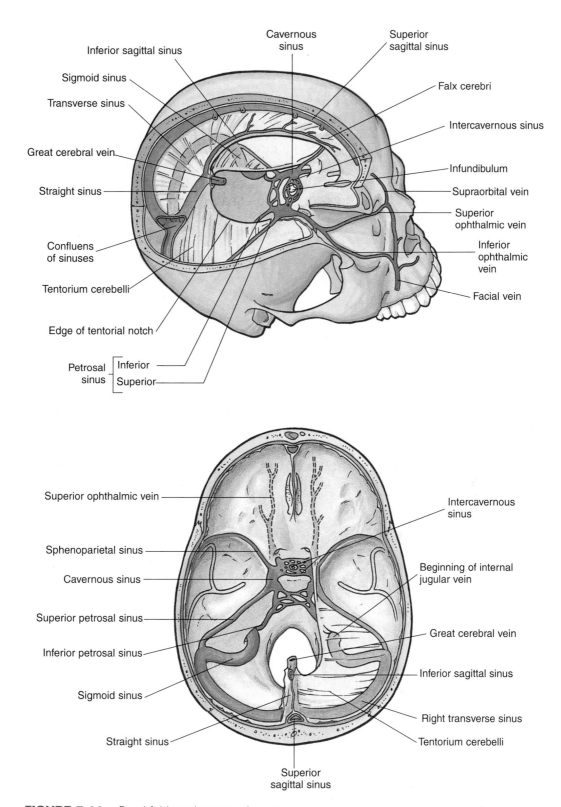

FIGURE 7.11. Dural folds and venous sinuses.

6. Dural Folds. Three dural folds exist and project inwardly to incompletely partition the cranial cavity. The **tentorium cerebelli** (L., tent) covers the cerebellum and lies between it and the overlying cerebral hemispheres. When standing upright, the tentorium supports the weight of the occipital lobes of the brain. The **falx cerebelli** (L., sickle) lies between the cerebellar hemispheres. The **falx cerebri** lies between the cerebral hemispheres and is attached to the **crista galli** anteriorly and fuses with the tentorium posteriorly. Its superior border encases the superior sagittal sinus. On its inferior border lies the **inferior sagittal dural sinus,** which drains into the **straight sinus.** The straight sinus then drains into the **confluens of the sinuses** (Fig. 7.11).   A5. 1:40:19-1:47:54/ G7.19-7.21/ C767/ R87/ N96-98

D. Removal of the Brain

1. With the cadaver in the prone position, cut the spinal cord at the level of the atlas (first cervical vertebra), and cut the paired vertebral arteries between the foramen magnum and the transverse process of the atlas.

2. Using the procedure described below, you will preserve the dural folds within the skull but still permit the removal of the brain. Your lab instructor will help you if necessary—don't be shy about asking for help! Turn the cadaver into the supine position and, while your lab partner supports the weight of the cerebral hemispheres, cut the falx cerebri close to the crista galli. Pull the falx superiorly and posteriorly. Now free the **olfactory bulbs** from the cribriform plate, cut the **infundibulum** just posterior to the **optic chiasma,** and sever the **optic nerves** and **two internal carotid arteries** close to the optic foramina. Lift the brain further and cut the **oculomotor nerves.** Detach the tentorium cerebelli on both sides by starting your cut at the free border of the tentorial notch and carry your cut posteriorly close to the superior margin of the petrous portion of the temporal bone. Next identify and cut the two **abducent nerves** and each of the remaining pairs of cranial nerves before gently "delivering" the brain from the cranial cavity.   G7.23/ C777/ R95/ N7, 98

E. Gross Examination of the Brain (always wear gloves)

1. On the cerebral hemispheres identify the following features:
A5. 11:01-40:13/ G7.24/ C778/ R89-90, 96/ N99

a. Frontal lobes

b. Temporal lobes

c. Occipital lobes

   d. **Parietal lobes**
   e. **Lateral sulcus** (fissure of Sylvius)
   f. **Central sulcus** (fissure of Rolando)

2. Now examine the base of the brain and identify the following arteries (Fig. 7.12):   **G7.34/ C775/ R94-95/ N132-136**

   a. **Vertebral arteries.** Enter the cranial cavity via the foramen magnum.
   b. **Posterior inferior cerebellar arteries.** Arise from the vertebrals.
   c. **Basilar artery**
   d. **Anterior inferior cerebellar arteries**
   e. **Superior cerebellar arteries.** Note that the oculomotor nerve (N. III) emerges between this artery and the posterior cerebral artery.
   f. **Posterior cerebral arteries**
   g. **Posterior communicating arteries**
   h. **Internal carotid arteries**
   i. **Cerebral arterial circle** (circle of Willis)

FIGURE 7.12.   Base of brain showing cerebral arterial circle and cranial nerves.

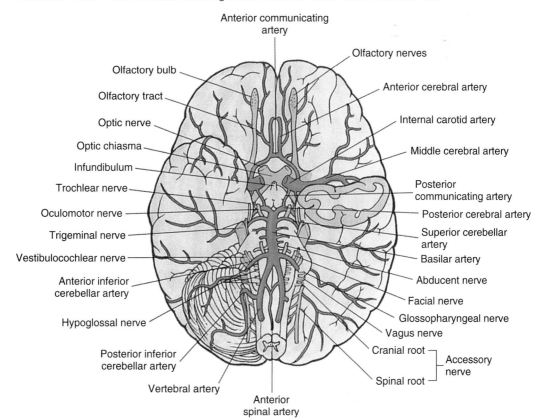

j. **Middle cerebral artery**

k. **Anterior cerebral arteries**

l. **Anterior communicating artery**

3. From the atlas and textbook, be sure you understand the relative distribution of blood supply to the brain by these arteries.

4. Also, on the base of the brain, identify the **twelve pairs of cranial nerves** (Fig. 7.12). Note that cranial nerves I and II are really brain tracts rather than peripheral nerves. Also note that the trochlear nerve is the only cranial nerve to arise from the dorsal aspect of the brain stem. This nerve will be hard to find on some brains! Begin now to appreciate the function, in general terms, of each of the 12 pairs of cranial nerves by reading about them in your textbook. **G7.22/ C778/ R68/ N108, 112**

F. Cranial Fossae

1. Anterior Fossa. Refer to a bony skull and note that this fossa is formed by the **sphenoid, ethmoid** and **frontal bones.** Note where cranial nerves leave the cranial fossae via foramen (Fig. 7.13). **G7.23/ C777/ R95/N98**

FIGURE 7.13. Interior of skull. Cavernous sinus is dissected on left side. Tentorium cerebelli also removed on the left side.

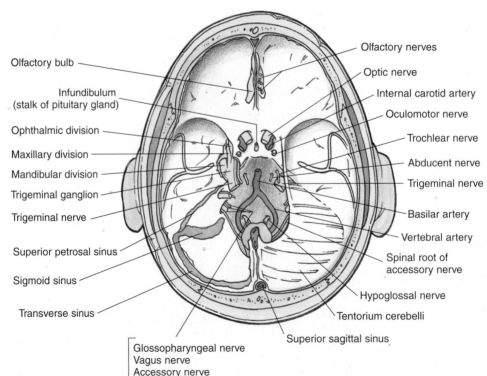

2. Middle Fossa. Primarily formed by the **sphenoid** and **temporal bones.** Note the **hypophyseal fossa** which contains the **pituitary gland.** Carefully remove the dura of the lateral wall of the **cavernous sinus** and identify the **abducent nerve** (CN. VI), **oculomotor nerve** (CN. III), **trochlear nerve** (CN. IV), **trigeminal nerve** (CN. V, notably, its first two divisions) and the **internal carotid artery** within the sinus. By stripping the dura from the sinus wall expose the **trigeminal ganglion** (Gasserian or semilunar ganglion) and the roots of the **three divisions of CN.V** (CN. V1: **ophthalmic division**; CN. V2: **maxillary division**; CN. V3: **mandibular division**). Learn where these nerves pass through foramen to exit the skull and read about the clinical importance of the cavernous sinus region. Finally, note the connections of the superior and inferior petrosal sinuses with the cavernous sinus.   **G7.39/ C769-774/ R137, 144/ N98**

3. Posterior Fossa. This fossa contains the **pons** and **medulla** of the brain as well as the occipital lobes of the cerebral hemispheres and the cerebellum. Note the location of the **internal acoustic meatus, jugular foramen** and **hypoglossal canal,** and learn which cranial nerves exit via these foramen (Fig. 7.13).   **G7.23/ C777/ R95/ N98**

V. ORBIT AND EYE

Learning Objectives

- Identify the key features of the living eye by examining your own eye in a mirror.
- List the bones which comprise the orbit.
- Describe how you would test each of the extraocular muscles and identify the nerve innervating each muscle.
- Compare and contrast the parasympathetic and sympathetic innervation to the eye and orbital region.
- Diagram how infections of the face might access the cavernous sinus via ophthalmic veins.
- Describe how the lacrimal gland is innervated by parasympathetics.
- Identify the three layers of the globe.

Key Concepts

- Circulation of aqueous humor in the eye
- Definition of the "special senses" from general senses
- Ophthalmic venous drainage
- Ophthalmic division of CNV
- Extraocular muscle testing

A. Bony Landmarks

1. Refer to the skull and identify: **A5. 2:00:40-2:03:51/ G7.42/ C787-788/ R128/ N1-2**

   a. **Maxillary bone**

   b. **Zygomatic bone**

   c. **Frontal bone**

   d. **Lacrimal bone**

   e. **Ethmoid bone**

   f. **Sphenoid bone**

   g. **Palatine bone**

   h. **Optic canal**

   i. **Superior orbital fissure.** Positioned between the **greater** and **lesser wings of the sphenoid bone.**

   j. **Inferior orbital fissure.** Gap between the maxilla and the greater wing of the sphenoid bone.

   k. **Infraorbital groove.** Continuous with the infraorbital canal and continuing anteriorly to the infraorbital foramen.

   l. **Anterior and posterior ethmoidal foramina and cribriform plate**

   m. **Lamina papyracea of the ethmoid bone**

2. On the right side, dissect the orbit from above and on the left side approach the orbit from the anterior (surgical) aspect. If necessary, inflate the partially collapsed eyeball by injecting water into the globe. Insert the needle obliquely through the transparent cornea in front of the pupil. The eyeball occupies the anterior half of the bony orbit and the posterior half is largely filled with fat and the extraocular muscles.

B. Right Orbit, Superior Approach

1. With a chisel or bone forceps, carefully remove the roof of the orbit as far anteriorly as possible but leave the superior orbital margin intact (Fig. 7.14A). The tough membrane just inferior to the bone is the **periorbita** which envelops the orbital contents. Expose the **frontal air sinus** and the **anterior** and **posterior ethmoid air cells (sinus)** medially, and note the mucous lining of the sinuses. Posteriorly, remove the lesser wing of the sphenoid bone and anterior clinoid process to expose the **superior orbital fissure** and **optic canal.** **G7.45/ C796, 803/ R136-137/N76, 80**

2. Nerves and Muscles (Fig. 7.14A). First, inject the eyeball with water to partially fill the collapsed globe. Now incise the pe-

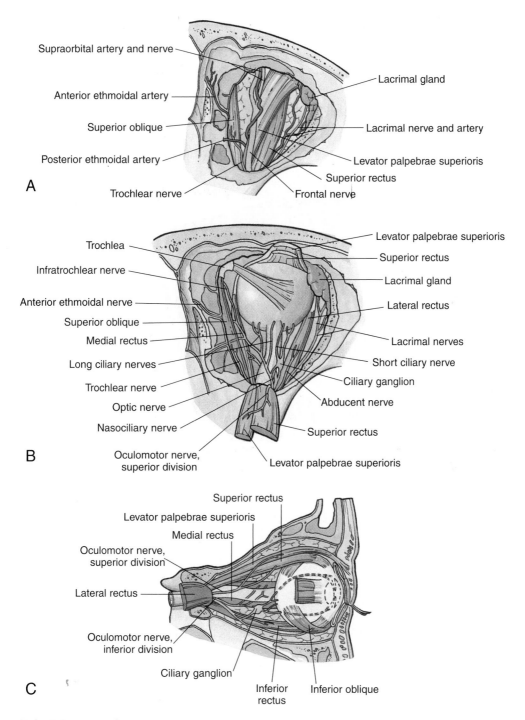

FIGURE 7.14. Orbital dissections.

riorbita and remove it. Locate the stump of the **trochlear nerve** and follow it anteriorly to the superior border of the **superior oblique muscle** which it innervates. Trace the large **frontal nerve** (from CN. V1) anteriorly and note its division into the **supratrochlear** and **supraorbital nerves.** (Remember:

CN. V1, 2, or 3 refers to the ophthalmic, maxillary or mandibular divisions, respectively, of CN. V). At this stage, remove the portion of the frontal bone forming the superior orbital margin to extend the dissection field. This exposes the upper eyelid and the **levator palpebrae superioris muscle** (Fig. 7.14B).   **A5. 2:03:52-2:13:12/ G7.45/ C803-808/ R136-137/ N81**

3. Next find the delicate **lacrimal nerve** (branch of CN. V1) and trace it anterolaterally to the **lacrimal gland.** Cut the levator palpebrae muscle as far anteriorly as possible and reflect it posteriorly. This reveals the underlying **superior rectus muscle.** Also cut this muscle anteriorly near where it attaches to the globe and reflect it posteriorly (Fig. 7.14B). A branch of the **oculomotor nerve** (CN. III) passes into the muscle on its underside. Examine the superior oblique muscle and anteriorly note its trochlea (pulley). Laterally, identify the **lateral rectus muscle** and look for its **abducent nerve** (CN. VI) on its medial side. Find the **nasociliary nerve** (from CN. V1) and try to identify its long ciliary branches. Another small branch of the nasociliary is the anterior ethmoidal nerve, which passes through the anterior ethmoidal foramen and ends on the nose as the external nasal nerve.   **G7.45-7.46/ C803-808/ R136-137/ N79, 81, 115**

4. Finally, locate the **oculomotor nerve** and note that it divides into superior and inferior branches. Attempt to find the very small **ciliary ganglion,** which lies just anterior to the apex of the orbit and lateral to the optic nerve. Short ciliary nerves connect it to the eyeball. From your textbook, understand the importance of this parasympathetic ganglion. Gently displace the globe to one side or the other and try to identify the **inferior rectus** and **inferior oblique muscles** (Fig. 7.14C).   **G7.46/ C806/ R137/N81, 115**

5. **Vessels** (Fig. 7.15). Find the **superior ophthalmic vein** and note that it anastomoses with tributaries of the facial vein at the medial angle of the eye. Posteriorly, the vein drains into the cavernous sinus. Inferiorly, small veins (not visible) connect the inferior ophthalmic vein with veins in the infratemporal fossa (pterygoid plexus of veins). Identify the **ophthalmic artery.** It arises from the **internal carotid artery,** and traverses the optic canal to enter the orbit inferior to the **optic nerve.** Remove the dural sheath around the optic nerve, slice the optic nerve in two and note the dark spot in the middle, which represents the central artery of the retina.   **G7.53/ C805-806/ R130/ N80, 86**

C. **Left Orbit, Surgical Approach** (Optional Dissection)

1. Before beginning, inject the eyeball with water to partially fill the collapsed globe and review the extent of the **conjunctiva.**

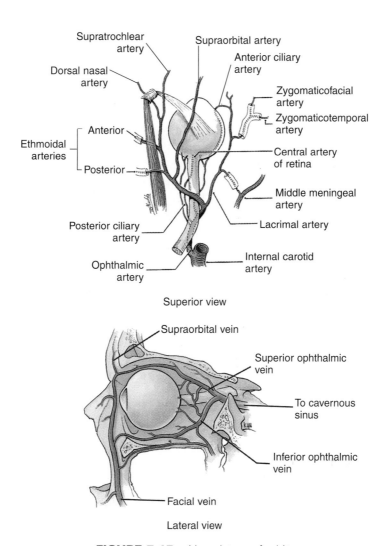

Superior view

Lateral view

FIGURE 7.15. Vasculature of orbit.

Then make a circular incision through the conjunctiva (not the sclera) about 6-8 mm from the sclerocorneal junction. Gently push a probe through this incision and locate one of the rectus muscles. To facilitate the dissection, remove both eyelids and the orbital septum. Now observe the four insertion points of the **rectus muscles** and study the **superior** and **inferior obliques.** **G7.46-7.49/ C807-816/ R131-132/ N76-79**

2. With a probe, hook each of the four rectus muscle tendons and cut across them near the globe. Insert a long scissors into the lateral side of the orbit and cut the **optic nerve.** Pull the globe anterior and sever the two oblique muscles and remove the eyeball. Identify the muscles and the common tendinous origin of the rectus muscles, the **anulus tendineus.** Between the two heads of the lateral rectus muscle, identify the ab-

ducens nerve, superior and inferior divisions of CN. III and
the nasociliary nerve.   G7.47/ C815-816/ R132/N79, 115

D. Dissection of the Globe

1. Remove the globe and with a scalpel cut the globe into two
   halves along a sagittal plane (Fig. 7.16). Carefully remove the
   remains of the vitreous. The eyeball consists of three layers: an
   external fibrous layer, a middle vascular pigmented layer, and
   an internal neural layer (really an extension of the central nervous system). Note the fibrous coat consisting of the **sclera** and

FIGURE 7.16.   Eyeball and ciliary region. *A,* anterior chamber; *P,* posterior chamber.

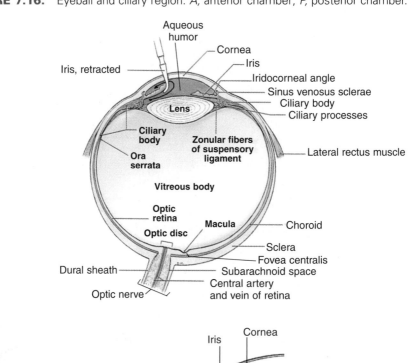

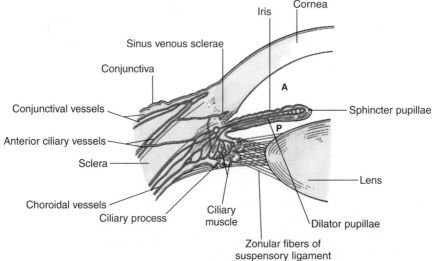

cornea. The middle or vascular coat consists of the **choroid, ciliary body** and **iris.** The internal or **retinal layer** is gray in color in the cadaver and may be partially detached from the underlying vascular coat (dark brown due to clotted blood). Note the **optic papilla** or disc. Identify the **lens** and the hole in the iris called the **pupil.**   A5. 2:03:52-2:13:12/ G7.51/ C817-823/ R129/ N82-86

2. Although not dissected, study the anatomy of the eyelids from your atlas and textbook, and examine your own eyelids in the mirror (wash your hands and be gentle!) Also, read about the lacrimal system and note on a bony skull the **lacrimal fossa** (contains the lacrimal sac) and the **nasolacrimal duct** which drains tears into the nasal cavity.   A5. 2:13:13-2:18:41/ G7.44/ C783-784/ R138/N76-77

VI.   PAROTID REGION

Learning Objectives

- Describe the innervation of the parotid gland by parasympathetic fibers.
- Describe the general sensory and general motor distribution of the facial nerve.

Key Concepts

- Terminal part of N. VII
- Parasympathetic fibers distribute on branches of CN. V

A. Bony Landmarks

1. Refer to the bony skull and identify: A4. 1:26:20-1:32:25/ G7.60-7.61/ C754, 868-869/ R25/ N9-10

   a. Temporal bone. Note the **styloid process, mastoid process, external acoustic meatus** and **mandibular fossa.**

   b. **Mandible.** Identify the **head, neck** and **ramus.**

   c. **Stylomastoid foramen.** The facial nerve (CN. VII) exits the skull here.

B. Muscles and Nerves (Fig. 7.17A)

1. First, observe the superficial structures which include the **parotid duct** (Stenson's duct), the **transverse facial artery** and **branches of the facial nerve.**   A4. 1:32:25-1:38:55, A4. 2:00:35-2:07:10/ G7.54/ C735/ R77/ N17, 19

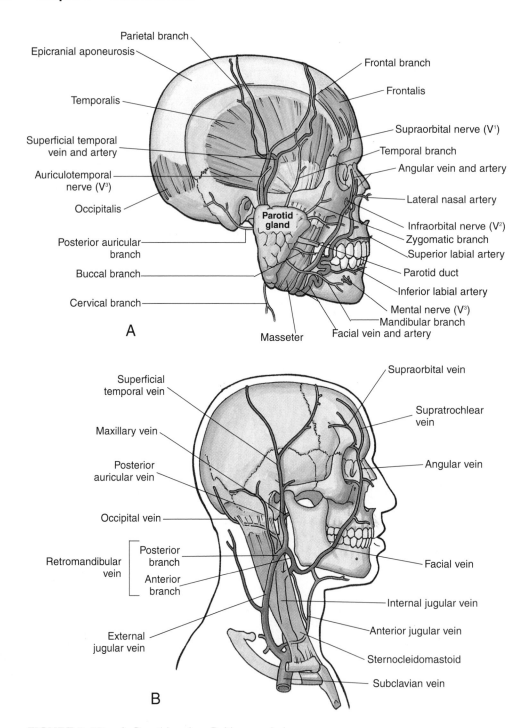

FIGURE 7.17. A, Parotid region. B, Venous drainage.

2. Begin the dissection by finding the **facial nerve** where it exits the stylomastoid foramen. Reflect the sternocleidomastoid muscle from the mastoid process. Then, retract the parotid gland anteriorly and hold it with a hemostat. With a probe, find the facial nerve exiting the stylomastoid foramen be-

tween the mastoid process and styloid process. Trace the facial nerve through the parotid gland using blunt dissection and attempt to identify representative fibers of the five terminal branches of the facial nerve supplying the muscles of facial expression: **temporal, zygomatic, buccal, mandibular** and **cervical** (Fig. 7.8).   **G7.10/ C736/ R77/ N19, 117**

3. Locate the **retromandibular vein** and **external carotid artery** deep to the gland. Posterior to the mandibular neck, the external carotid terminates as the superficial temporal and maxillary arteries (we will dissect these later).

4. Examine the **posterior belly of the digastric muscle** and **stylohyoid.** Look for and clean the **auriculotemporal nerve** (again, we will dissect later), a branch of CN. V, which carries parasympathetic postganglionic fibers to the parotid gland. This nerve may be difficult to identify at this time.   **G7.54/ C748/ R77/ N29, 35**

VII. TEMPORAL REGION

Learning Objectives

- Name the muscles of mastication and describe their function.
- Describe the distribution of the maxillary artery and list its main branches.
- Describe the connections of the pterygoid plexus of veins and state why they are important.
- On a diagram, label the major branches of the mandibular division of the trigeminal.
- Describe the movements at the temporomandibular joint (TMJ).

Key Concepts

- Muscles of mastication
- Maxillary artery's three parts
- Pterygoid venous plexus
- Mandibular division of CN. V

A. Introduction. The temporal region consists of the temporal fossa (superior to the zygomatic arch) and the infratemporal fossa (inferior and deep to the arch). The lateral wall of the infratemporal fossa is the ramus of the mandible. The infratemporal fossa contains the muscles of mastication, the mandibular division of CN. V, and the maxillary artery.

B. Bony Landmarks

1. On the bony skull identify the following features: G7.60-7.61/
   C754-755, 781/ R28-31/N2, 5, 11

   a. **Temporal fossa.** Formed by components of four bones: parietal, frontal, squamous part of the temporal bone, and the greater wing of the sphenoid bone.

   b. **Zygomatic arch**

   c. **Mandible.** Note the **ramus** and **angle.** Identify the **mandibular notch** between the **head** and **coronoid process.** On the inner aspect, find the **lingula, mandibular foramen** and **mylohyoid line.**

   d. **Lateral pterygoid plate of sphenoid**

   e. **Infratemporal surface of maxilla**

   f. **Pterygopalatine fossa.** Wedge-shaped space between the lateral pterygoid plate and maxilla.

   g. **Greater wing of the sphenoid**

   h. **Foramen ovale and spinosum.** In the greater wing of the sphenoid bone.

C. Masseter Muscle

1. Clean the **masseter muscle** (Table 7.4) and detach the posterior third from the zygomatic arch. Reflect the detached part anteriorly, noting the masseteric nerve and vessels supplying this muscle (difficult to find).   **A4. 1:32:25-1:38:55/ G7.55/ C748/ R58-59/ N48**

2. Now pass a probe deep to the zygomatic arch to protect deep structures and make the saw cuts shown in Figure 7.18 to remove the zygomatic arch. Reflect the bony arch and attached masseter inferiorly removing its deep attachment to the ramus of the mandible (sever masseteric nerve and vessels). Leave the masseter attached only to the lower margin of the mandible.

3. Next review the **temporalis muscle.** Note its attachments and find a small fatpad between the temporalis and lateral orbital wall. It is continuous with the buccal fatpad.   **G7.55/ C749/ R58/ N48**

D. Infratemporal Fossa

1. This fossa contains two muscles of mastication, the mandibular division of the trigeminal nerve (CN. V3), and the maxillary vessels.   **A4. 1:29:40-1:32:25**

**TABLE 7.4
MUSCLES ACTING ON TEMPOROMANDIBULAR JOINT**

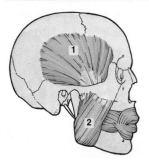 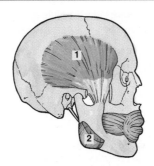 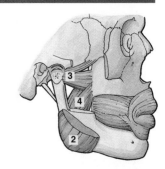

Muscle	Origin	Insertion	Innervation	Main Actions
Temporalis (1)	Floor of temporal fossa and deep surface of temporal fascia	Tip and medial surface of coronoid process and anterior border of ramus of mandible	Deep temporal branches of mandibular n. (CN V³)	Elevates mandible, closing jaws its posterior fibers retrude mandible after protrusion
Masseter (2)	Inferior border and medial surface of zygomatic arch	Lateral surface of ramus of mandible and its coronoid process	Mandibular n. via masseteric nerve that enters its deep surface	Elevates and protrudes mandible, thus closing jaws; deep fibers retrude it
Lateral pterygoid (3)	*Superior head:* infratemporal surface and infratemporal crest of greater wing of sphenoid bone *Inferior head:* lateral surface of lateral pterygoid plate	Neck of mandible, articular disc, and capsule of temporomandibular joint	Mandibular nerve (CN V³) via lateral pterygoid nerve from anterior trunk, which enters its deep surface	Acting together, they protrude mandible and depress chin; acting alone and alternately, they produce side-to-side movements of mandible
Medial pterygoid (4)	*Deep head:* medial surface of lateral pterygoid plate and pyramidal process of palatine bone *Superficial head:* tuberosity of maxilla	Medial surface of ramus of mandible, inferior to mandibular foramen	Mandibular nerve (CN V³) via medial pterygoid nerve	Helps to elevate mandible, closing jaws; acting together, they help to protrude mandible; acting alone, it protrudes side of jaw; acting alternately, they produce a grinding motion

2. Make the saw cuts shown in Figure 7.18. We will remove the coronoid process first. To do this, push a probe through the mandibular notch and swing it obliquely inferiorly and anteriorly in close contact with the mandible. This protects structures deep to the coronoid process. Then cut the coronoid process obliquely and reflect it together with its attached temporalis muscle. Raise the temporalis and see its nerves and vessels on the underside.

3. Next we'll remove the upper part of the mandibular ramus. With a pencil, mark the approximate position of the lingula on the lateral surface of the mandible. Pass the blade of an open forceps medial to the neck of the mandible, keep close to the bone and move it inferiorly until it is arrested by the lingula. Cut horizontally across the ramus of the mandible as shown in Figure 7.18 and remove the bone fragment. Fi-

FIGURE 7.18. Dash lines indicate cuts to gain access to the infratemporal fossa.

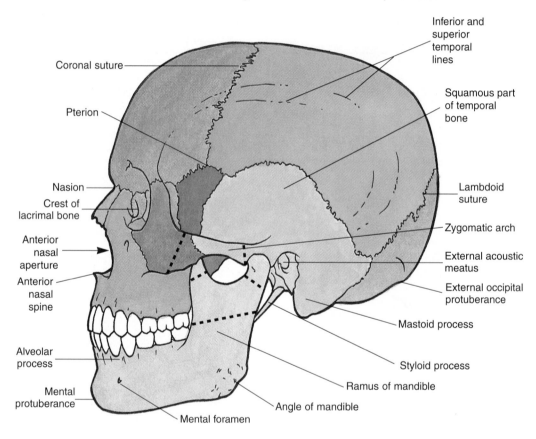

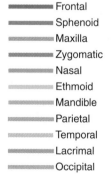

Frontal

Sphenoid

Maxilla

Zygomatic

Nasal

Ethmoid

Mandible

Parietal

Temporal

Lacrimal

Occipital

nally, cut through the neck of the mandible as shown in Figure 7.18 just inferior to the **temporomandibular (TMJ) joint.**

4. Identify the **inferior alveolar nerve** and **artery** (Fig. 7.19). Remove any mandibular periosteum that may obscure your view. Follow these structures into the mandibular canal. Note

FIGURE 7.19. Dissections of the infratemporal region. A, Superficial. B, Deep.

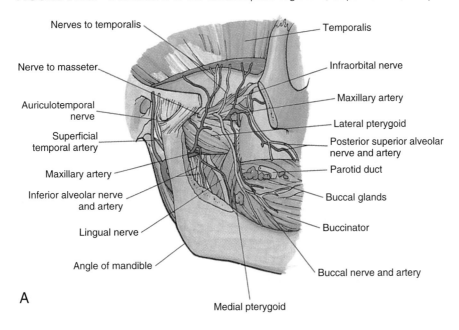

Nerves to temporalis
Temporalis
Nerve to masseter
Infraorbital nerve
Auriculotemporal nerve
Maxillary artery
Superficial temporal artery
Lateral pterygoid
Maxillary artery
Posterior superior alveolar nerve and artery
Inferior alveolar nerve and artery
Parotid duct
Lingual nerve
Buccal glands
Angle of mandible
Buccinator
Buccal nerve and artery
Medial pterygoid

A

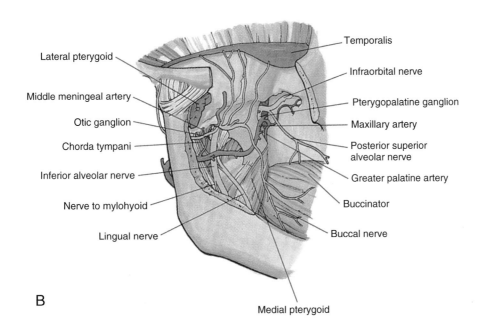

Lateral pterygoid
Temporalis
Middle meningeal artery
Infraorbital nerve
Otic ganglion
Pterygopalatine ganglion
Chorda tympani
Maxillary artery
Inferior alveolar nerve
Posterior superior alveolar nerve
Nerve to mylohyoid
Greater palatine artery
Lingual nerve
Buccinator
Buccal nerve
Medial pterygoid

B

the delicate **mylohyoid nerve** leaving the inferior alveolar just before it passes into the bony canal. A ligament, the sphenomandibular ligament, may still be attached to the lingula. If so, don't mistake it for the inferior alveolar nerve. Cut the ligament if it is in your way.   **A5. 40:33-1:03:31/ G7.62/ C750/ R65/ N41, 65**

5. Next find the **lingual nerve** which runs anterior and superior to the inferior alveolar nerve. It emerges from the inferior border of the **lateral pterygoid muscle** (Fig. 7.19A).   **G7.62-7.63/ C750/ R64, 72-73/ N41, 65**

6. To follow the **maxillary artery** it is necessary to remove the **lateral pterygoid muscle.** The muscle has two heads: one arises from the roof of the fossa and the other from lateral pterygoid plate. Sever the muscle's attachments and remove it in a piecemeal fashion. Now identify the small **chorda tympani nerve** which joins the lingual nerve proximally (Fig. 7.19B). This nerve brings taste and parasympathetic fibers from CN. VII, which distribute with branches of the lingual nerve, a branch of CN. V3. Follow the inferior alveolar and lingual nerves back to the foramen ovale. Look in the middle cranial fossa to find this foramen and the main trunk of CN. V3. Find the **buccal nerve** (purely sensory to the cheek mucosa) and again find the **auriculotemporal nerve.** Usually, the nerve loops around the **middle meningeal artery.**   **G7.63/ C751/ R72-73, 80/ N41, 63**

7. Identify and clean the **inferior alveolar artery, maxillary artery** and look for the **middle meningeal artery** passing through the **foramen spinosum.** Viewing the foramen spinosum from the middle cranial fossa, stick a pin through the foramen to locate the vessel. The maxillary artery (Fig. 7.19B) also gives off muscular branches to the muscles of mastication and a **posterior superior alveolar artery** to the maxillary molar teeth. Other branches of the maxillary will be studied later but read in your textbook about its three parts.   **G7.62/ C746, 750/ R65/ N41, 63-65**

E. Temporomandibular Joint (TMJ)

1. Although the superior portion of the mandibular ramus has been removed, the mandibular head and neck are still intact. Study the **joint capsule** and note it is thickened laterally to form the temporomandibular ligament. Open the joint cavity and note the **articular disc.** In the atlas and textbook, study the attachments of the lateral pterygoid muscle and understand how the TMJ works.   **G7.58-7.59/ C737/ R55-57/ N11, 49**

VIII. RETROPHARYNGEAL REGION

Learning Objectives

■ Describe the craniovertebral articulation.
■ Define the "retropharyngeal" space.
■ Trace sympathetic nerve fibers from the upper thoracic cord to the superior cervical ganglion, naming each nerve traversed.
■ List the muscles innervated by the IXth, Xth, XIth, and XIIth cranial nerves.

Key Concepts

■ Retropharyngeal or "danger space"
■ Structures exiting the jugular foramen

A. Introduction. In this dissection, the head and visceral neck (prevertebral region) will be detached from the vertebral column and prevertebral muscles. The logical plane for this dissection is the **retropharyngeal (retrovisceral) space** which extends from the base of the skull and passes inferiorly into the superior mediastinum. Identify this space on a cross-section of the neck.   **A4. 2:10:30-2:20:15/ G8.2-8.28/ C890/ R161/ N30**

B. Bony Landmarks

1. On a skull and vertebral column identify the following: **A4. 00:55-26:40/ G4.39/ C639-644/ R28, 184// N12-13**

a. **Axis.** Note the **dens** or **odontoid process.**

b. **Atlas.** Identify the **posterior arch, anterior arch, transverse process** and **superior articular facet.**

c. **Occipital bone.** Note the **foramen magnum** and **occipital condyles.**

C. Craniovertebral Joints

1. Insert the fingers of both hands into the space posterior to the carotid sheaths and visceral neck structures. Work your fingers superiorly to the base of the skull and inferiorly to T3, freeing up any fascial attachments. On both sides, reflect the two sternocleidomastoid muscles.

2. With the cadaver in the prone position, remove all of the musculature on the back of the neck down to the atlas. A wedge-shaped piece of the occipital bone was removed previously. Now, remove the **posterior arch of the atlas** to expose the dura. Excise the dura, remove any spinal cord and expose the **tectorial membrane** (Fig. 7.20). Cut this membrane transversely about 1 cm superior to the anterior border of the foramen magnum, raise the membrane, and reflect it inferiorly as far as possible.

3. Now identify the craniovertebral ligaments: the **transverse ligament of the atlas** which holds the dens firmly to the anterior arch of the atlas. This ligament and the vertically oriented superior and inferior bands are collectively known as the **cruciform ligament** (Fig. 7.20). Next, identify the **alar ligaments** (check ligaments) that extend from the dens to the lateral margins of the foramen magnum. They are about the thickness of a probe. Cut these ligaments and note how the head rotates easily.   **G4.39-4.40/ C649-652/R191-192/ N15**

[handwritten note in left margin: Cutting alar ligaments / allows head to rotate]

D. Removal of the Head

1. Cut along the border of the foramen magnum to sever any attachments. Force a chisel into the atlanto-occipital joint and disarticulate the joint (Fig. 7.20). Roll the cadaver over into the supine position and pull the cervical viscera and large vessels anteriorly. Sever the sympathetic trunk on one side superior to the superior cervical ganglion. Now pass a knife between the transverse process of the atlas and the occipital bone to sever the prevertebral muscles attaching to the skull. Carry your knife cut across the anterior arch of the atlas to sever the anterior atlanto-occipital membrane. Do not cut the cranial nerves IX, X, XI, and

FIGURE 7.20.   Posterior view of craniovertebral joints. The tectorial membrane is cut, revealing the cruciform ligament.

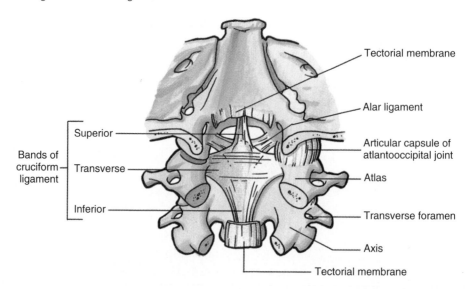

XII. Now detach the head and cervical viscera by bringing them forward and placing the chin of the cadaver on its own chest. Seek help from an instructor if you encounter difficulty.

E. Pre- and Lateral Vertebral Regions

1. Examine the prevertebral fascia investing the prevertebral muscles (Table 7.5). On the side where the sympathetic trunk was left in place, identify the **superior, middle** and **inferior**

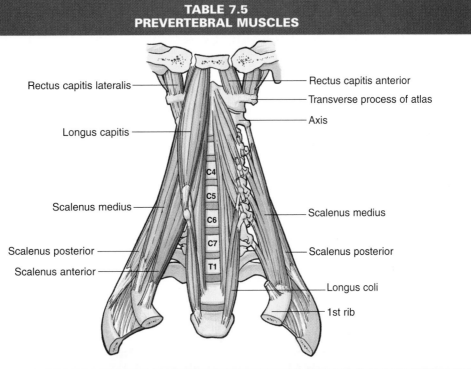

TABLE 7.5
PREVERTEBRAL MUSCLES

Rectus capitis lateralis
Rectus capitis anterior
Transverse process of atlas
Axis
Longus capitis
Scalenus medius
Scalenus medius
Scalenus posterior
Scalenus posterior
Scalenus anterior
Longus coli
1st rib

Muscle	Superior Attachment	Inferior Attachment	Innervation	Main Actions
Longus colli	Anterior tubercle of C1 vertebra	Body of T3 vertebra with attachments to bodies of C1–C3 and transverse processes of C3–C6 vertebrae	Ventral rami of C2–C6 spinal nn.	Flexes cervical vertebrae
Longus capitis	Basilar part of occipital bone	Anterior tubercles of C3–C6 transverse processes	Ventral rami of C2–C3 spinal nn.	Flexes head
Rectus capitis anterior	Base of skull, just anterior to occipital condyle	Anterior surface of lateral mass of C1 vertebra (atlas)	Branches from loop between C1 and C2 spinal nn.	Flexes head
Rectus capitis lateralis	Jugular process of occipital bone	Transverse process of C1 vertebra (atlas)		Flexes head and helps to stabilize the head

cervical ganglia (not always present). Observe **rami communicantes.** G8.33/ C716/ R160-161/ N65, 124-125

2. Identify the **longus colli, capitis** and **scalenus anterior muscles.** Remove portions of these muscles from the transverse
processes and note the underlying spinal nerves. Follow the
vertebral artery into the transverse foramen of C6 (variable).
G8.33/ C717/ R62/ N14, 25, 28

F. Base of the Skull

1. Looking at the back of the head and visceral neck, observe
where the internal jugular vein leaves the skull and identify
CN. IX, X and **XI** (Fig. 7.21). Find CN. XII. A5. 1:05:10-1:21:04/
G8.36/ C892/ R160-161/ N65, 124

FIGURE 7.21. Pharynx. Posterior view showing related nerves and vasculature.

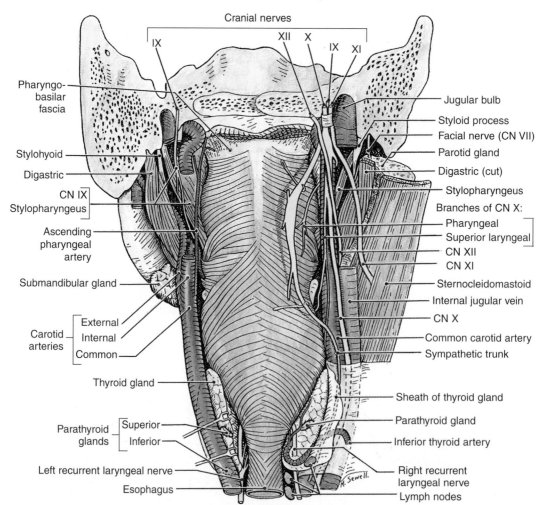

Posterior view

2. Follow the length of **CN. X** and try to identify the **inferior ganglion** (nodose) just below the jugular foramen. Locate the **superior laryngeal nerve.**

3. Trace the **CN. XI** where it lies lateral to the vagus and passes through the interval between the internal jugular vein and internal carotid artery. CN. XI crosses anterior to the internal jugular vein in about 70% of all cases.

4. Follow the pathway of **CN. XII.** Note that it loops laterally around the carotid sheath structures.

5. Finally, find **CN. IX.** It is best seen just under the investing fascia covering the **stylopharyngeus muscle** on its posterior or lateral side.   **G8.36/ C892/ R160-161/N65, 124**

6. Review these cranial nerves and their distribution and function in your textbook. You should know how to clinically test each of the cranial nerves.

IX. PHARYNX

Learning Objectives

- Describe the actions of the pharyngeal constrictor muscles in swallowing.
- Name the three descriptive regions of the pharynx.
- List the lymphoid tissue that comprises Waldeyer's tonsillar ring.
- Compare and contrast the motor and sensory innervation of the superior laryngeal and recurrent laryngeal nerves.

Key Concepts

- Development of this region from the pharyngeal arch system
- Muscles involved in swallowing
- Waldeyer's tonsillar ring

A. Introduction. The pharynx is the superior end of the respiratory and digestive tubes. It extends from the base of the skull to the inferior border of the cricoid cartilage (vertebral level C6).

The pharynx is covered by an areolar coat of fascia that is continuous with the fascia covering the buccinator, so it is called the **buccopharyngeal fascia.** A pharyngeal plexus of nerves and vessels lies beneath this layer. Next, the pharyngeal wall consists of a muscular layer and then a fibrous layer called the pharyngobasilar fascia, which anchors the pharynx to the base of the skull. The inner wall of the pharynx consists of a submucosa and mucous membrane coat.   G8.36/ C888-890/ R162-163/N54

B. External Pharynx

1. Clean the posterior aspect of the pharynx to reveal the three pharyngeal constrictors (Fig. 7.21). The **middle constrictor** is easiest to identify as its fibers arise from the hyoid bone and stylohyoid ligament (Table 7.6). Above it lies the **superior constrictor** which arises from the **pterygomandibular raphe** and lies behind the mandible. The **inferior constrictor** arises from the thyroid and cricoid cartilages. Note that the three constrictor muscles overlap one another like shingles on a roof.   A4. 2:10:30-2:20:15/ G8.36/ C891/ R163/ N61, 69

2. Between the middle and inferior constrictors find the **internal laryngeal branch of the superior laryngeal nerve** piercing the **thyrohyoid membrane.** Inferior to the inferior constrictor, find the **recurrent laryngeal nerves** on each side. Between the middle and superior constrictors, find the **stylopharyngeus muscle** and CN. IX.   G8.36/ C892/ R158, 160-161/ N65, 70

C. Internal Pharynx

1. Slit open the posterior pharyngeal wall starting at the cricoid cartilage and carry the incision along the midline to the base of the skull. Internally, the pharynx is divided into three descriptive regions (Fig. 7.22):

a. Nasal pharynx. Identify the posterior nasal apertures (spaces) the **nasal choanae,** separated by the **nasal septum.** On each side of the nasal pharynx, 1 to 1.5 cm posterior to the inferior concha, is the opening for the **auditory (Eustachian) tube.** Cartilage bulging around the opening of the tube is called the **torus tubarius.** This opening may be difficult to visualize at this time but you'll see it later. G8.38/ C893-895/ R159/ N60-61

b. Oral pharynx. Identify the **palatoglossal arch** and **palatopharyngeal arch,** noting that the **palatine tonsil** lies in the bed formed between these two arches (Table 7.7). Examine these same arches in the mouth of your lab partner. G8.37-8.38/ C893-895/ R159/ N58, 60-61

c. Laryngeal pharynx or hypopharynx. Examine the **epiglottis** and identify the **piriform recesses** (small bones or food particles may become trapped in this space during swallowing). Carefully remove the mucosa covering the piriform recess and expose the **internal laryngeal nerve** (from

TABLE 7.6
MUSCLES OF PHARYNX

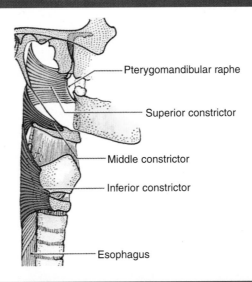

Pterygomandibular raphe

Superior constrictor

Middle constrictor

Inferior constrictor

Esophagus

Muscle	Origin	Insertion	Innervation	Main Action(s)
Superior constrictor	Pterygoid hamulus, pterygomandibular raphe, posterior end of mylohyoid line of mandible, and side of tongue	Median raphe of pharynx and pharyngeal tubercle	Pharyngeal and superior laryngeal branches of vagus (CN X) through pharyngeal plexus	Constrict wall of pharynx during swallowing
Middle constrictor	Stylohyoid ligament and superior (greater) and inferior (lesser) horns of hyoid bone	Median raphe of pharynx		
Inferior constrictor	Oblique line of thyroid cartilage and side of cricoid cartilage			
Palatopharyngeus	Hard palate and palatine aponeurosis	Posterior border of lamina of thyroid cartilage and side of pharynx and esophagus		Elevate pharynx and larynx during swallowing and speaking
Salpingopharyngeus	Cartilaginous part of auditory tube	Blends with palatopharyngeus		
Stylopharyngeus	Styloid process of temporal bone	Posterior and superior borders of thyroid cartilage with palatopharyngeus	Glossopharyngeal n. (CN IX)	

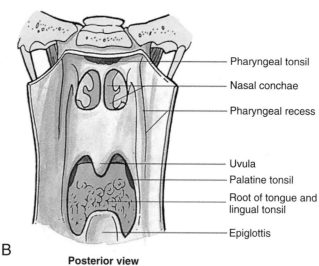

Pharynobasilar fascia

Nasal septum

Auditory tube

Levator veli palatini

Tensor veli palatini

Styloid process

Hamulus of medial pterygoid plate

Stylopharyngeus

Superior constrictor

Salpingopharyngeus

Middle constrictor

Palatopharyngeus

Musculus uvulae

Root of tongue

Inferior constrictor

Epiglottis

Aryepiglottic fold

Oblique and transverse arytenoid

Posterior cricoarytenoid

Esophagus

A

Posterior view

Pharyngeal tonsil

Nasal conchae

Pharyngeal recess

Uvula

Palatine tonsil

Root of tongue and lingual tonsil

Epiglottis

B

Posterior view

FIGURE 7.22. A. Muscles of the soft palate and interior of the pharynx. The posterior pharyngeal wall has been cut in the midline and reflected laterally, and the mucosa and submucosa removed to show the muscles. B. Interior of the pharynx showing pharyngeal recess and tonsils.

superior laryngeal nerve) and the **recurrent laryngeal nerve** (see Fig. 7.26). Note their sensory distribution to the hypopharynx in your textbook.   G8.50/ C898/ R158-159/ N60-61, 70

2. Although difficult to see in the cadaver, note that collections of lymphoid tissue around the opening of the auditory tube (tubal tonsil), in the roof of the nasopharyngeal region (pharyngeal tonsil), in the palatine fossa between the palatoglossal and palatopharyngeal folds (palatine tonsil), and at the base of the tongue (**lingual tonsil**) form a "protective" ring of lymphoid tissue called Waldeyer's tonsillar ring. With age, these tonsillar structures usually atrophy.

X. BISECTION OF THE HEAD

Learning Objectives

- Compare and contrast the tongue's innervation via five cranial nerves.
- Name where the lacrimal duct and paranasal sinuses open in the nasal cavity.
- Identify the bones that contribute to the nasal septum and lateral nasal wall.
- Describe the distribution of the maxillary division of the trigeminal nerve.

Key Concepts

- Tongue innervation
- Distribution of sphenopalatine artery
- CN. V2 and the pterygopalatine ganglion

A. Introduction

1. To gain access to the nasal and oral cavities, it will be necessary to bisect the head. Usually the nasal septum is deviated to one side or the other and we will bisect the head just laterally to the deviated septum on the side with the wider nasal passage.

2. In the midline, start by cutting through the upper lip and nose until you reach the nasal bone. Then, keeping a saw blade close to the septum, saw through the nasal and frontal bones, and extend your cut through the cribriform plate, body of the sphenoid bone, dorsum sellae and basioccipital bone to the foramen magnum. Finally, with a scalpel, divide the uvula

and the soft palate. Now, the two superior halves of the head will fall apart from each other, and the tongue lies exposed.

B. Tongue

1. Note the following features on the cadaver's tongue (Fig. 7.23). Also, study these features on your own tongue by looking in a mirror: **A4. 1:45:15-1:50:50/ G7.68/ C860/ R145/ N52**

   a. **Sulcus terminalis** dividing the anterior two thirds from the posterior one third of the tongue.

FIGURE 7.23. Features of the tongue (top) and its innervation (bottom).

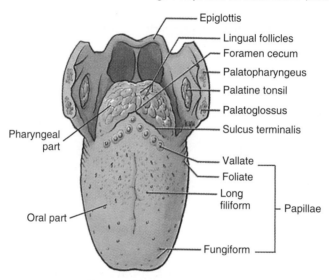

Dorsum of tongue

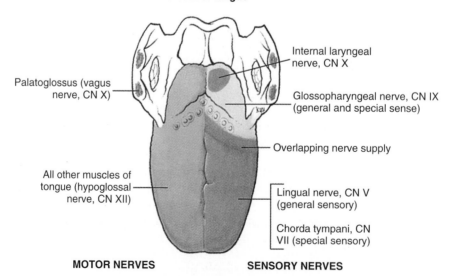

Innervation of tongue

b. **Foramen cecum.** Marks the site of the embryonic thyroglossal duct that was attached to the developing thyroid gland.

c. **Fungiform papillae.** Mushroom-shaped papillae that appear as red spots on your tongue.

d. **Filiform papillae.** Sensitive to touch.

e. **Vallate papillae.** Just anterior to the sulcus terminalis.

f. **Median glossoepiglottic fold** running from the tongue to the epiglottis and flanked on either side by the **valleculae.**

g. **Lingual tonsils.** Lymphoid follicles over the posterior third of the tongue.

2. Now, bisect the mandible and tongue: split the median raphe of the **mylohyoids** to separate the muscles, divide the **geniohyoids** and then saw the mandible in the midsagittal plane exactly between the geniohyoid muscles. Finally, bisect the tongue in the midsagittal plane from the tip to the epiglottis, but do not cut the epiglottis or the hyoid bone. Be aware of the fact that the tongue is supplied by five different cranial nerves (CN. V, VII, IX, X, XII) (Fig. 7.23).

C. Nasal Cavities

1. On a bony skull identify the following features: A4. 58:25-1:12:35/ G7.85-7.86/ C827-829/ R45, 48/ N3, 5, 33

   a. **Palatine process of the maxilla.**

   b. **Horizontal plates of the palatine bones.**

   c. **Incisive foramen.**

   d. **Greater and lesser palatine foramina.**

   e. **Perpendicular plate of the palatine bone.**

   f. **Sphenopalatine (pterygopalatine) foramen.**

   g. **Pterygopalatine fossa.**

2. Identify the following features on a bisected skull: G7.85/ C827-828/ R46, 48/ N33-34

   a. **Sphenoid sinus.** Right and left may vary in size.

   b. Posterior opening for the **pterygoid canal.** Often the canal forms a ridge in the floor of the sphenoid sinus and connects the **foramen lacerum** with the **sphenopalatine foramen.**

   c. **Foramen lacerum.**

   d. **Sphenopalatine foramen.**

   e. **Frontal process of the maxilla.**

    f. **Inferior concha (turbinate).**

    g. **Middle and superior conchae,** part of the ethmoid bone.

    h. **Maxillary sinus.**

    i. **Nasolacrimal canal.**

    j. **Frontal sinus.**

    k. **Cribriform plate.**

    l. **Nasal septum** made up of the unpaired **vomer, perpendicular plate of the ethmoid bone** and a **septal cartilage.**

3. Strip the mucoperiosteum completely off of the nasal septum. Next, carefully remove the bony and cartilaginous portions of the septum but leave the lateral mucoperiosteum. In this mucoperiosteum, note the presence of the septal vessels but do not attempt to dissect them. Trace the **nasopalatine nerve** (branch of CN. V2) to the **incisive canal.** Now, remove the remaining portions of the septum and mucoperiosteum to expose the lateral nasal wall. **G7.86/ C830/ R139/ N35, 38**

D. Lateral Wall of the Nasal Cavity

1. Identify the **superior, middle** and **inferior conchae** (Fig. 7.24). Identify the opening of the **auditory tube** from the middle ear. **A4. 1:12:35-1:23:40/ G7.83/ C832/ R140/ N32**

2. With scissors, cut away the inferior concha and identify the

FIGURE 7.24. Lateral nasal wall. Parts of the conchae are cut away to reveal openings. A rod placed in the frontal sinus shows the opening into the middle meatus, and a rod in the sphenoid sinus shows its opening into the sphenoethmoidal recess.

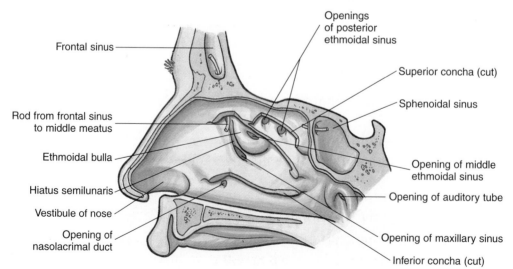

nasolacrimal duct opening into the **inferior meatus.** Also cut away the middle concha and identify the **hiatus semilunaris.** Note the **ostium for the maxillary sinus.** This hiatus has a sharp lower edge and a rounded superior edge formed by the **ethmoidal bulla,** an elevation of the ethmoidal sinuses. Identify the **frontal sinus** and its opening into the **infundibulum,** a narrow passage anterosuperior to the hiatus semilunaris. Staying above the hiatus semilunaris, break into the middle and anterior ethmoid air cells, and remove the superior concha to reveal the posterior ethmoid air cells. Also, locate the **sphenoid sinus** posteriorly.   G7.83/ C831-832/ R141-143/ N32

3. With a bone forceps, remove the medial wall of the maxillary sinus to reveal a ridge of bone in the roof of the sinus which is the **infraorbital canal** containing the **infraorbital nerve** and **vessels.** Break into the canal to see these structures. In the floor of the sinus, look for roots of the maxillary teeth.   G7.67/ C835-836/ R83/ N40, 42-43

E. Pterygopalatine Ganglion

1. Strip the mucoperiosteum from the perpendicular plate of the palatine bone and insert a needle into the greater palatine canal just medial to the third maxillary molar tooth. Leaving the needle in place, use a probe to break away the medial wall of the greater palatine canal to expose the **greater palatine nerve** and **vessels.**   G7.67, 7.75/ C834-836/ R143/ N35-38

2. Follow the greater palatine nerve superiorly to the **sphenopalatine foramen** and locate the large **pterygopalatine ganglion.** The ganglion and **sphenopalatine artery** lie in the sphenopalatine foramen and are situated directly behind the middle turbinate (concha).   G7.85/C833-834/ R143/ N37-39

3. Next, locate the **nerve of the pterygoid canal** as follows: remove the mucoperiosteum of the sphenoid sinus and look for a ridge of bone in the floor produced by the pterygoid canal. Open this bony ridge to reveal the nerve and follow the nerve (Vidian nerve) into the ganglion. Read about the autonomic nerves passing to the pterygopalatine ganglion in your textbook.

4. Review in the atlas and textbook the distribution of the sphenopalatine artery and its very small branches to the nasal septum and lateral nasal wall. Also, review the course and distribution of the maxillary division of CN. V.

XI. PALATE, MOUTH, AND NASOPHARYNGEAL WALL

Learning Objectives

- Describe how the muscles of the tongue, palate, and pharynx function during swallowing.
- Describe the innervation of the submandibular and sublingual salivary glands.
- Identify key anatomical features in your own mouth.

Key Concepts

- Deglutition (swallowing)
- Parasympathetic innervation of salivary glands

A. Palatine Region

1. If present, remove the **palatine tonsil** from its bed and, dissecting through the thin pharyngobasilar fascia, expose the palatopharyngeus and superior constrictor muscles. Again, identify the **stylopharyngeus muscle** and the **glossopharyngeal nerve,** and bluntly dissect and demonstrate that the glossopharyngeal nerve spreads out to the mucosa covering the posterior third of the tongue. Identify the **styloglossus muscle,** extending from the styloid process to the lateral aspect of the tongue (Fig. 7.25).   G8.45/ C843-846/ R147/ N58-59, 61

B. Sublingual Region

1. Outside of lab, using your clean index finger, explore the vestibule of your mouth and palpate the following structures: A4. 1:26:20-1:29:40, A4. 1:52:30-1:58:30/ G8.43/ C843-846/ R149, 151/ N45, 55

   a. Maxilla. Its anterior and infratemporal aspects.

   b. Inferior border of the **zygomatic arch.**

   c. Ramus of the mandible.

   d. Coronoid process and tendon of the temporalis muscle.

   e. Masseter muscle. Clench your teeth to feel the muscle.

   f. Frenulum (L., bridle) of upper and lower lips.

   g. Orifice of the parotid duct. Opposite the 2nd maxillary molar tooth.

   h. Frenulum linguae. Connects the tongue to the floor of the mouth.

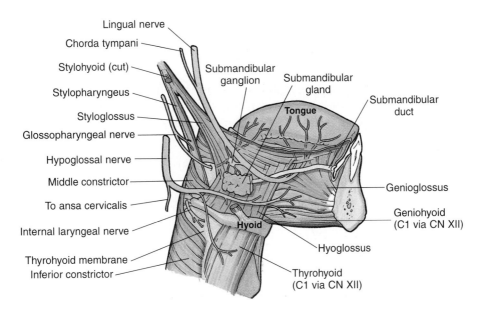

FIGURE 7.25. Tongue muscles and nerves.

i. **Deep lingual veins.**

j. **Opening of the submandibular duct.**

k. **Plica sublingualis.** Several small ducts of the sublingual gland open onto this plica.

l. **Hamulus of the medial pterygoid plate.**

2. On a bony mandible identify the mylohyoid line (attachment of mylohyoid muscle) and **sublingual fossa** (for sublingual salivary gland). On the bisected head, identify the **mylohyoid, geniohyoid** and **genioglossus muscles.** G7.57/ C867-869/ R55/ N53, 57

3. Carefully incise the mucous membrane along the furrow between the plica sublingualis and the tongue. Start at the frenulum of the tongue and carry your incision back to the 2nd mandibular molar tooth. Displace the sublingual gland laterally and the tongue medially and look for small ducts coming from the gland and opening on the summit of the plica. Identify the **submandibular salivary gland** and **submandibular duct** running medially along the **sublingual gland.** Find the **lingual nerve** posterior to the last molar tooth and trace it anteriorly (Fig. 7.25). It spirals around the submandibular duct as follows: first lateral and superior to the duct, then inferior and finally inferior and medial to the duct where it divides into several branches that enter the tongue (Fig. 7.25). G7.70/ C855-858/ R148-149/ N41, 53, 55

4. Identify the **submandibular ganglion** in the vicinity of the 3rd molar tooth, attached to the lingual nerve by several short branches. **G8.18/ C855/ R148/ N53**

5. Find the **hypoglossal nerve** where it lies between the submandibular gland and **hyoglossus muscle,** inferior to the lingual nerve. **G8.17-8.19/ C855/ R147/N53-54**

C. Tongue

1. Approaching the bisected head from the lateral aspect, detach the mylohyoid muscle from the hyoid bone and expose the **hyoglossus muscle** (Table 7.8). The hypoglossal and lingual nerves cross the hyoglossus (Fig. 7.25). Locate the **lingual artery** and follow it medial (deep) to the hyoglossus by reflecting the muscle superiorly after cutting its attachment to the hyoid bone. **A4. 1:38:55-1:50:50/ G8.20/ C854, 897/ R146-147/ N47, 53**

2. On one side of the cadaver only, make a transverse section through the tongue to note its intrinsic musculature (consists of vertical, transverse and longitudinal fibers).

TABLE 7.7 **MUSCLES OF SOFT PALATE**				
Muscle	Superior Attachment	Inferior Attachment	Innervation	Main Action(s)
Levator veli palatini	Cartilage of auditory tube and petrous part of temporal bone	Palatine aponeurosis	Pharyngeal branch of vagus n. via pharyngeal plexus	Elevates soft palate during swallowing and yawning
Tensor veli palatini	Scaphoid fossa of medial pterygoid plate, spine of sphenoid bone, and cartilage of auditory tube		Medial pterygoid n. (a branch of mandibular n.)	Tenses soft palate and opens mouth of auditory tube during swallowing and yawning
Palatoglossus	Palatine aponeurosis	Side of tongue	Cranial part of CN XI through pharyngeal branch of vagus n. (CN X) via pharyngeal plexus	Elevates posterior part of tongue and draws soft palate onto tongue
Palatopharyngeus	Hard palate and palatine aponeurosis	Lateral wall of pharynx		Tenses soft palate and pulls walls of pharynx superiorly, anteriorly, and medially during swallowing
Musculus uvulae	Posterior nasal spine and palatine aponeurosis	Mucosa of uvula		Shortens uvula and pulls it superiorly

TABLE 7.8
EXTRINSIC MUSCLES OF TONGUE

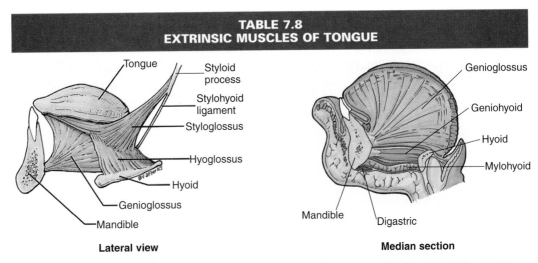

Lateral view

Median section

Muscle	Origin	Insertion	Innervation	Actions
Genioglossus	Superior part of mental spine of mandible	Dorsum of tongue and body of hyoid bone		Depresses tongue; its posterior part protrudes tongue
Hyoglossus	Body and greater horn of hyoid bone	Side and inferior aspect of tongue	Hypoglossal n. (CN XII)	Depresses and retracts tongue
Styloglossus	Styloid process and stylohyoid ligament			Retracts tongue and draws it up to create a trough for swallowing
Palatoglossus	Palatine aponeurosis of soft palate	Side of tongue	Cranial root of CN XI via pharyngeal branch of CN X and pharyngeal plexus	Elevates posterior part of tongue

D. Nasopharyngeal Wall. Removal of the mucosa of the palatine arches reveals the small **palatoglossus** and **palatopharyngeus muscles** comprising these folds (Table 7.7) (also see Fig. 7.22A). Palpate the hamulus of the medial pterygoid plate (just posterior to the third mandibular molar tooth) and realize that the fibrous **pterygomandibular raphe** lies between the hamulus and mandible. The **superior constrictor** posteriorly and the **buccinator muscle** anteriorly meet at this raphe.   G8.40/ C852, 896/ R147/ N49, 54, 58-59

1. **Auditory Tube.** Identify the **opening of the Eustachian tube** and appreciate that the anterior two thirds of the tube is cartilaginous while the inferior and lateral walls of the tube are membranous.   G8.39/ C846/ R141/ N58-59, 93

2. **Levator and Tensor Veli Palatini Muscles.** Just anterior to the ostium of the auditory tube incise the mucosa and identify the thick (pencil size) **levator veli palatini (levator palati) mus-**

cle, and anterior and laterally (deeper from this aspect) the **tensor veli palatini (tensor palati) muscle.** Understand their important actions. The tensor muscle arises from the scaphoid fossa and winds around the **hamulus** to insert into the soft palate. This action allows the muscle to tense the soft palate while the levator elevates the palate (see Fig. 7.22A). **G8.40/ C888, 895/ R64/ N46, 58-59, 61**

XII. LARYNX

Learning Objectives

- List which muscles adduct or tense the vocal folds and which muscles abduct the folds.
- Identify the cartilages that comprise the larynx.
- Describe the motor and sensory innervation of the larynx by the superior and recurrent laryngeal nerves.

Key Concepts

- Action of the vocal folds to produce phonation
- Superior and recurrent (inferior) laryngeal nerves

A. Laryngeal Cartilages

1. From your atlas and available models, study the cartilages that comprise the laryngeal skeleton. These include the **cricoid** (Gk., ring), **thyroid,** arytenoid, corniculate and cuneiform cartilages. **A4. 2:25:00-2:28:15/ G8.46/ C899-900/ R154/ N71**

B. Laryngeal Muscles

1. Strip the mucosa from the entire pharyngeal aspect of the larynx to expose the **posterior cricoarytenoid muscles** and the **arytenoideus** (Figs. 7.22A and 7.26).

2. Cut through the ligaments that hold the cricothyroid joint together and reflect the thyroid lamina by cutting the lamina about 8 mm to the left of the midline. Reflect the lamina which now is attached to the cricoid cartilage by the **cricothyroid muscle.** Identify the following muscles: **A4. 2:30:45-2:35:30/ G8.52/ C904/ R156/ N72**

   a. Cricothyroid.

   b. Lateral cricoarytenoid.

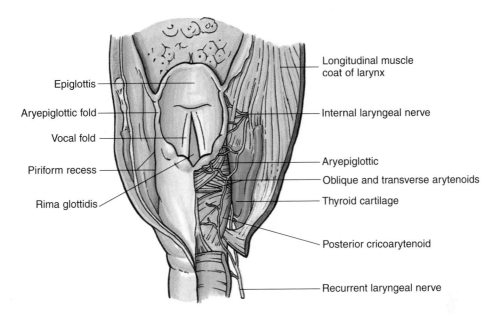

Epiglottis

Aryepiglottic fold

Vocal fold

Piriform recess

Rima glottidis

Longitudinal muscle
coat of larynx

Internal laryngeal nerve

Aryepiglottic

Oblique and transverse arytenoids

Thyroid cartilage

Posterior cricoarytenoid

Recurrent laryngeal nerve

FIGURE 7.26. Muscles and nerves of larynx. On the right side, the mucosa is removed to reveal muscles, nerves, and the thyroid cartilage.

    c. **Thyroarytenoid.**

    d. **Vocalis.** A small muscle in the lateral and inferior portion of the vocal ligament.

    e. **Thyroepiglotticus.**

3. Realize from your atlas, textbook or models that the posterior cricoarytenoid opens the **rima glottidis** (space between the vocal folds), or abducts the vocal cords, and is the only pair of muscles capable of this action.   **G8.50-8.58/ C904-909/ R156/ N73**

C. Interior of the Larynx

1. The larynx consists of three spaces: the **vestibule, ventricle** and **infraglottic cavity** (Fig. 7.27).

2. Examine the interior of the larynx from the superior aspect and note the **vestibular (false) folds** lying superolateral to the **vocal (true) folds** or **cords.** With a scissors, split the trachea, lamina of the cricoid, and arytenoid muscle in the posterior median plane and view the interior of the larynx. Remove the mucous membrane from one half of the interior of the larynx between the cricoid cartilage and vocal ligament. This reveals the **conus elasticus** or **cricothyroid ligament.** The thickened margin of the conus forms the **vocal ligament (fold).**   **A4.**
2:28:15-2:30:45/ G8.58/ C907-909/ R157/ N71-75

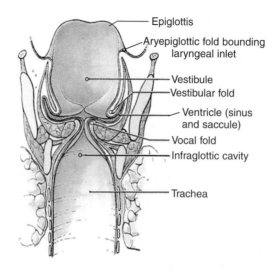

Coronal section

FIGURE 7.27.   Compartments of larynx. Coronal section viewed posteriorly.

3. Review the innervation of the larynx (motor and sensory) via the superior and recurrent laryngeal branches of the vagus nerve in your textbook (Fig. 7.26).

XIII.   EAR

Learning Objectives

- Identify the middle ear ossicles and describe the transduction mechanism that converts sound waves into nerve impulses in CN. VIII.
- Describe the course of the facial nerve through the temporal bone.
- Name the types of nerve fibers found in the greater petrosal and chorda tympani nerves.
- Identify the bony features of the inner ear which contribute to the vestibular apparatus.

Key Concepts

- Course of CN. VII
- Chorda tympani nerve
- Greater petrosal nerve
- Middle ear ossicles
- Vestibular system

A.   Introduction. Because the features of the ear are very small, be sure to read in your textbook about the important structures that comprise the auditory and vestibular systems. In this dissection, which may be optional in your course (check with your instructor), we will focus only on the larger features that can be easily observed without the aid of a dissecting microscope.

Complete this dissection on only one side of the skull. The features of the middle and inner ear are contained within the **temporal bone**, which must be carefully dissected. Some structures will be lost or damaged, even by those of you who are careful, so feel free to look at other dissections in the laboratory. The faculty may have temporal bone "models" that will help you with your orientation and identification.

If decalcified temporal bones are available, make the cuts described in this dissection with a sharp scalpel blade and carefully "shave" the bone away to reveal the inner and middle ear structures. If only the cadaver specimen is available, carefully chisel the bone as described.

B. Bony Landmarks

1. Refer to a skull and identify the following features: A5.
   2:20:50-2:40:48/ G7.2/ C919-920/ R121, 126/ N93

   a. **Mastoid process.**
   b. **External acoustic meatus.**
   c. **Internal acoustic meatus.**
   d. **Tegmen tympani.** A plate of the petrous part of the temporal bone.
   e. **Jugular fossa and foramen.**
   f. **Auditory tube.**
   g. **Carotid canal.**
   h. **Stylomastoid foramen.**
   i. **Styloid process.**

C. Inner Ear

1. Strip the dura from the middle cranial fossa and make a parasagittal saw cut through the petrous portion of the temporal bone just lateral to the **internal acoustic meatus,** as shown in Figure 7.28. Extend the cut deeply to the floor of the middle cranial fossa. Now, begin to carefully chisel or chip away the tegmen tympani working laterally from the saw cut (direction of the arrows in the figure). Hold the chisel parallel to the floor of the middle cranial fossa.

2. Following along the internal acoustic meatus, identify the **facial nerve** and **geniculate ganglion,** and look for the anteri-

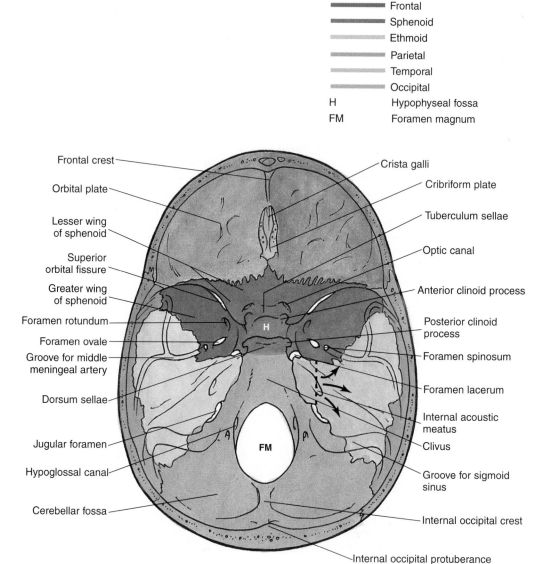

Frontal
Sphenoid
Ethmoid
Parietal
Temporal
Occipital
H   Hypophyseal fossa
FM   Foramen magnum

Frontal crest
Orbital plate
Lesser wing of sphenoid
Superior orbital fissure
Greater wing of sphenoid
Foramen rotundum
Foramen ovale
Groove for middle meningeal artery
Dorsum sellae
Jugular foramen
Hypoglossal canal
Cerebellar fossa

Crista galli
Cribriform plate
Tuberculum sellae
Optic canal
Anterior clinoid process
Posterior clinoid process
Foramen spinosum
Foramen lacerum
Internal acoustic meatus
Clivus
Groove for sigmoid sinus
Internal occipital crest
Internal occipital protuberance

FIGURE 7.28. Dashed line is saw cut and arrows show direction of chisel cuts on right side.

orly running greater petrosal nerve. **G7.99/ C929-938/ R74, 119-123/ N89, 92**

3. Continue to chip bone away anterior to cranial nerves VII and VIII, and identify the **cochlea.** Chip bone away posteriorly to reveal one or more of the **semicircular canals** and **vestibule** (have a lab instructor show you these on a temporal bone model) **G7.94, 7.104-7.105/ C938-939/ R123-126/ N90, 92**

D. Middle Ear

1. The middle ear (tympanic cavity) is a small rectangular box-shaped portion of the ear that contains the middle ear ossicles (Fig. 7.29).

Continue to carefully chip bone from the tegmen tympani laterally in the direction of the external acoustic meatus. In so doing, you will eventually remove the bony roof of the middle ear and reveal the **malleus, incus** and **stapes.** The head of the malleus usually is the most easily identified of the middle ear ossicles. On the lateral wall, note the **tympanic membrane** (ear drum) and the **chorda tympani nerve** passing medially to the handle of the malleus and tympanic membrane. Anteriorly, in the roof of the auditory tube, try to locate the tensor tympani muscle (inserts into the malleus).   **A5. 2:33:07-2:37:58/ G7.100-7.101/ C921-931/ R118-124/ N88-89**

2. By chipping bone posteriorly you will expose the **mastoid air cells** lying posterior to the middle ear cavity.

3. Review the course of the facial nerve through the temporal bone and appreciate that the mucosa of the middle ear (sensory fibers) is innervated by the tympanic branch of CN. IX. **G7.99-7.101/ C931-937/ R74, 123/ N89, 92, 117-118**

FIGURE 7.29.   Walls of tympanic cavity. Viewed from the anterior aspect as though looking up through the auditory tube into the middle ear cavity.

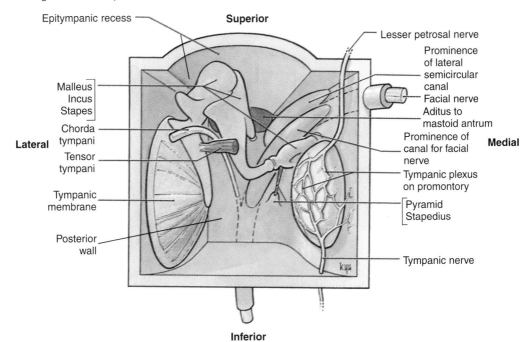

INDEX

Page numbers in italics indicate figures. Page numbers followed by "t" indicates tables.